ALLE ZEIT WACH
1842

Sigmund Pomer William E. Hull (Eds.)

Magnetic Resonance in Nephrourology

Clinical and Experimental Applications

With 77 Figures and 7 Tables

Springer-Verlag
Berlin Heidelberg New York
London Paris Tokyo
Hong Kong Barcelona
Budapest

PD Dr. med. Sigmund Pomer
Chirurgische Universitätsklinik
Abteilung Urologie und Poliklinik
Im Neuenheimer Feld 110
D-69120 Heidelberg, FRG

William E. Hull, PhD
German Cancer Research Center
Central Spectroscopy Department
Im Neuenheimer Feld 280
D-69120 Heidelberg, FRG

ISBN-13:978-3-642-78068-4

Library of Congress Cataloging-in-Publication Data. Magnetic resonance in nephrourology: clinical and experimental applications/S. Pomer, W.E. Hull (eds.). p. cm. Includes index.
ISBN-13:978-3-642-78068-4 e-ISBN-13:978-3-642-78066-0
DOI: 10.1007/978-3-642-78066-0 1. Urinary organs—Magnetic resonance imaging. 2. Kidneys—Magnetic resonance imaging. 3. Kidneys—Cancer—Diagnosis. 4. Organ Preservation. I. Pomer, S. (Sigmund) II. Hull, W. E. (William E.) [DNLM: 1. Nuclear Magnetic Resonance. 2. Kidney—physiology. 3. Kidney Neoplasms—physiopathaology. WJ 301 M196 1993] RC874.M34 1993 616.6'07548—dc20
DNLM/DLC for Library of Congress 93-9946 CIP

Typesetting: Best-set Typesetter Ltd., Hong Kong
21/3130-5 4 3 2 1 0 – Printed on acid-free paper

Contents

List of Contributors

Addresses are given at the beginning of the respective contribution.

Magnetic Resonance Spectroscopy and Tumor Cell Metabolism

Introduction to Magnetic Resonance and Its Possible Applications in Nephrourology: High-Field Spectroscopic Characterization of Renal Cell Carcinoma

W.E. Hull[1]

Introduction

The purpose of this introductory chapter is to provide some basic information regarding the fundamentals of magnetic resonance (MR) for those readers who are less familiar with the technique. Emphasis here is on MR spectroscopy and its possible applications in research and clinical practice, hopefully stimulating the reader to look more closely at the various topics covered by the contributors to this volume. Finally, some applications of high-field ^{13}C and ^{31}P MR spectroscopy in the area of renal cell carcinoma are briefly illustrated.

The Magnetic Resonance Phenomenon

Many natural nonradioactive isotopes have a nuclear spin (a magnetic moment) and can be induced to give radiofrequency MR signals in a strong, applied magnetic field (B_0). Important isotopes for biological MR applications re listed in Table 1 along with their abundance, resonance frequency (which is proportional to B_0) and relative detection sensitivity. It is important to note that the time needed to perform an MR measurement, either spectroscopy or imaging, achieving a desired signal-to-noise (S/N) or quality level, is proportional to the inverse square of the isotope sensitivity or the spin (substance) concentration. Each factor of two improvement in S/N for a given measurement requires a factor of 4 increase in measurement time. On the other hand, for a given substance concentration the measurement time required decreases approximately with the third power of B_0, so that high-field in vitro studies have a considerable advantage over low-field clinical in vivo studies in terms of detection threshold. The abundant isotope of carbon ^{12}C has no nuclear spin; thus, the low sensitivity for ^{13}C means that labeling of substances is often necessary for biological applications.

[1] Central Spectroscopy Department, German Cancer Research Center, Im Neuenheimer Feld 280, D-69120 Heidelberg, FRG

Table 1. Principal isotopes used for biological applications of MR

Isotope	Natural abundance (%)	Resonance frequency at 11.7 T (MHz)	Detection sensitivity[a]	Minimal concentration[b] (nmol/g)
^{1}H	100	500	100	4
^{19}F	100	470	83	5
^{31}P	100	202	7	50
^{13}C	1.1	125	0.02	4000 (40[c])

[a] In relative units for equal molar concentrations of substance.
[b] Minimum metabolite detection threshold for a 30-min measurement at 11.7 T.
[c] For >90% ^{13}C labeling.

Since MR operates at radiofrequency energies rather than at optical, X-ray or γ-ray energies, it is a basically harmless but very insensitive technique compared to other spectroscopic or radiological methods. However, this is the price that we pay for the incomparably detailed molecular, biochemical, and physiological information that MR can provide.

The exact resonance frequencies for a given isotope at various positions in a molecule are sensitive to the local molecular structure. These effects, called "chemical shifts," are proportional to B_0 and are in the Hz to KHz range. By convention, chemical shifts are usually expressed in units of ppm relative to the resonance frequency of a chosen reference signal so that the values are independent of B_0.

It is important to note that: (a) an MR spectrum is a fingerprint for chemical structure, i.e., the position and form of the signals can be used to identify molecular species; (b) mixtures can be analyzed without chromatographic separation; the area under a signal (its integral) is proportional to the concentration of the substance; (c) signal linewidths contain information about molecular mobility and chemical exchange rates; and (d) higher magnetic fields improve signal separation (resolution) and detection sensitivity. It is very important to realize that signal linewidths are roughly proportional to molecular weight; *macromolecules and tightly bound metabolites are not normally detected* or at best are seen as a very broad background signal. It is of particular interest that the MR technique is truly nondestructive and can be applied in a noninvasive manner in vivo.

There are a variety of ways to generate and manipulate MR signals to provide: (a) purely chemical/structural information (spectroscopy); (b) purely spatial information (imaging); and (c) both chemical and spatial information (localized spectroscopy). The MR techniques used and their applications (spectroscopy and imaging) are generally classified as in vitro or in vivo. In vitro or "high-resolution" MR techniques are employed at field strengths of $B_0 = 2-14$ T to obtain detailed analytical information on:

- Homogeneous solutions or mixtures, cellular extracts
- Biofluids: urine, plasma, bile, CSF (typically 0.3–1.5 ml)

- Solid tissue, biopsies (0.1–1 g)
- Cell suspensions (10^6–10^8 cells)
- Perfused cells in a bioreactor or gel matrix
- Perfused, functioning organ or cold preserved organ (e.g., up to 7 T/21-cm magnet for human kidneys)

In vivo MR methods are more or less noninvasive but involve some compromises in sensitivity and resolution. Animal model studies can be performed at fields comparable to those for in vitro work, B_0 = 2–11.7 T, while clinical studies are limited to B_0 = 1–4 T. The techniques of interest are:

- Implanted coils or special coils for externalized kidneys in animal models (highest sensitivity)
- Surface coil techniques to study localized subcutaneous tumors, for example, human renal cell carcinoma (RCC) xenografts on nude mice
- More general "whole-body" imaging and image-guided spectroscopic techniques for internal organs (reduced sensitivity)

Some typical applications for the most important MR isotopes are the following:

^{1}H. Signals from a large number of molecules can be detected, i.e., metabolic fingerprinting; high-resolution imaging and MR microscopy using the intense signals from tissue water and fat; low-resolution imaging (0.1–10 cm^3 voxels, ranging from high-field animal studies to medium-field human studies) of specific metabolites that occur in m*M* concentrations.

^{13}C. Only a few metabolites at high concentation are detectable in natural abundance; labeled precursors can be used to trace metabolic pathways and transport; two-dimensional $^{1}H/^{13}C$ experiments can be used for metabolic fingerprinting.

^{19}F. High detection sensitivity; specific fluorine-containing analogs can be used for studying metabolism or transport, for example, fluorosugars; fluorine-containing antimetabolites, for example, fluoropyrimidines, are used for chemotherapy; perfluorocompounds in blood substitutes can be used for imaging with mm resolution.

^{31}P. For monitoring phosphate, phosphoester, and nucleotide levels to determine intracellular pH and energy status for judging organ viability, studying phospholipid metabolism in tumors and monitoring therapy response; metabolic imaging requires ca. 1–100 cm^3 voxels, depending on subject and field strength.

MR in Nephrourology

There are many possibilities for the application of MR spectroscopy, either purely as an analytical technique or for studying problems related to the biochemistry, physiology, and function of the kidney. Several of these topics are addressed in this volume:

- Detailed characterization and quantitation of individual metabolites that are important in kidney function and preservation
- Elucidation of enzyme mechanisms and metabolic pathways, transport mechanisms, and kinetics
- Investigation of processes involved in ischemic insult, conservation, and reperfusion
- Investigation of processes involved in malignant transformation and tumor growth
- Monitoring the effects of hormones, cytokines, drugs, and other cancer therapies

MR imaging has both its clinical aspects and, at high field strengths, can be used for research studies to obtain "microscopic" resolution [1] in animal

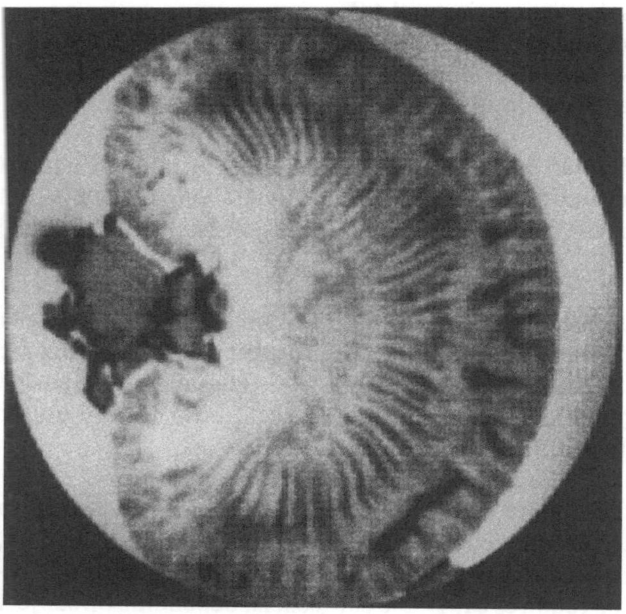

Fig. 1. 300-MHz (7-T) ^{1}H MR microscopy of a freshly excised rat kidney flushed with Eurocollins solution and stored at 10°C in a 20-mm MR tube. The 256 × 256 image matrix has a planar pixel resolution of 86 μm for a 250-μm XY slice along the long axis of the kidney, obtained using a gradient-echo method (TR = 60 ms, TE = 4.8 ms, flip angle = 30°, G_x = 0.107 mT/mm, G_z = 0.218 mT/mm) in 1 h measurement time with a Bruker AM-300 "super-wide-bore" spectrometer

models or ex vivo (e.g., see Fig. 1). Some of the possible applications are also treated in this volume:

- Characterization of morphological abnormalities and tissue damage due to ischemia and storage
- Determination of blood flow and real-time visualization of renal function using paramagnetic contrast agents
- Diagnosis of urological disorders
- Diagnosis of malignancy, monitoring of therapy
- MR microscopy of human renal cell carcinoma as xenografts on nude mice [6]

A reasonable strategy for MR research is to combine in vitro and in vivo techniques at the highest possible field strengths to obtain high-quality, detailed biochemical data which can provide a solid foundation for developing useful applications of in vivo MR under the technical and ethical limitations of the clinical environment.

High-Field ^{31}P MR Spectroscopy of Human Renal Cell Carcinoma

^{31}P MR is of particular interest for studying tumor cells since they can differ from normal cells in pH, energy metabolism, and phospholipid turnover (proliferation rate) [2, 11]. Human tumor cells can be studied as perfused cultures in an appropriate matrix [3–5, 8–10] or as solid tumors (xenografts) in the nude mouse [4, 5].

Methods. We have begun to study human RCC [5] with in vitro MR at 11.7 T (202 MHz). Typically $2-8 \times 10^7$ cultured cells are harvested and mixed 1:1 with basement membrane gel (BMG; Matrigel EHS, Serva) and extruded as 250 µm diameter threads which can be coiled into a compact form and perfused (1 ml/min) in a 10-mm MR sample tube. This allows: (a) complete control of medium, additives, pH, temperature, oxygen; and (b) continuous monitoring over several days with a time resolution of 30–60 min for individual spectra. RCC can also be studied as small solid tumors (subcutaneous or intramuscular) using in vivo MR. At 7 T (121 MHz) the quality of spectra is not quite as good as can be obtained from cultures [4, 5], but the major metabolites can be resolved with good sensitivity using a 10 to 15-mm surface coil for localized spectroscopy. Repetitive and noninvasive measurements, even without anesthetics, can be made to follow growth and therapy.

Figures 2 and 3 present some initial results with a renal cell carcinoma line established in the tumor bank of the German Cancer Research Center. Each peak in a spectrum represents a particular type or class of phos-

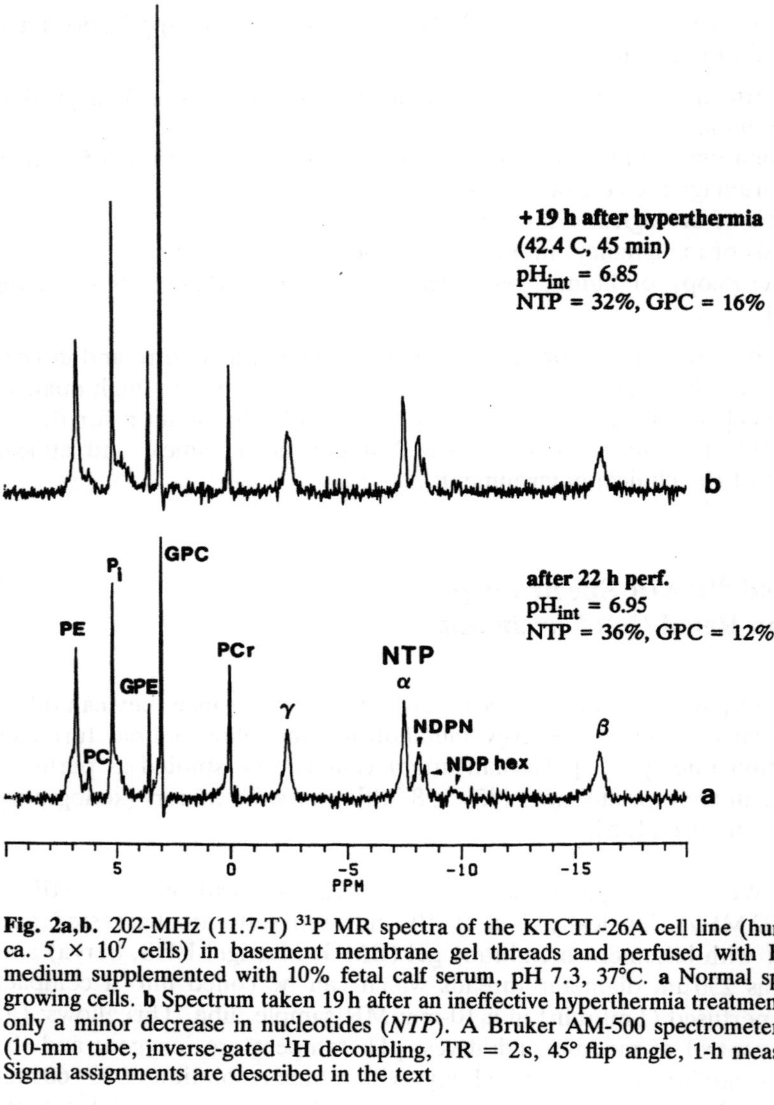

Fig. 2a,b. 202-MHz (11.7-T) ^{31}P MR spectra of the KTCTL-26A cell line (human RCC, ca. 5×10^7 cells) in basement membrane gel threads and perfused with Ham's F12 medium supplemented with 10% fetal calf serum, pH 7.3, 37°C. **a** Normal spectrum of growing cells. **b** Spectrum taken 19 h after an ineffective hyperthermia treatment, showing only a minor decrease in nucleotides (*NTP*). A Bruker AM-500 spectrometer was used (10-mm tube, inverse-gated ^1H decoupling, TR = 2 s, 45° flip angle, 1-h measurement). Signal assignments are described in the text

phorus atom with a specific molecular environment. The assignments are as follows: NTP α, β, γ are the three phosphate moieties for all nucleoside-5′-triphosphates (ATP, UTP, etc. cannot be separately resolved); NDPN represents the α, β phosphate groups for all dinucleotides of the NAD type; NDP-hex are all nucleotide hexoses such as UDPG (the α phosphate overlaps with NDPN; the β phosphate signal is at ca. −9.5 ppm); PCr is phosphocreatine (its chemical shift has been defined as 0 ppm), P_i is inorganic phosphate, either extracellular *e* (1 m*M* in medium) or intracellular *i*. Phos-

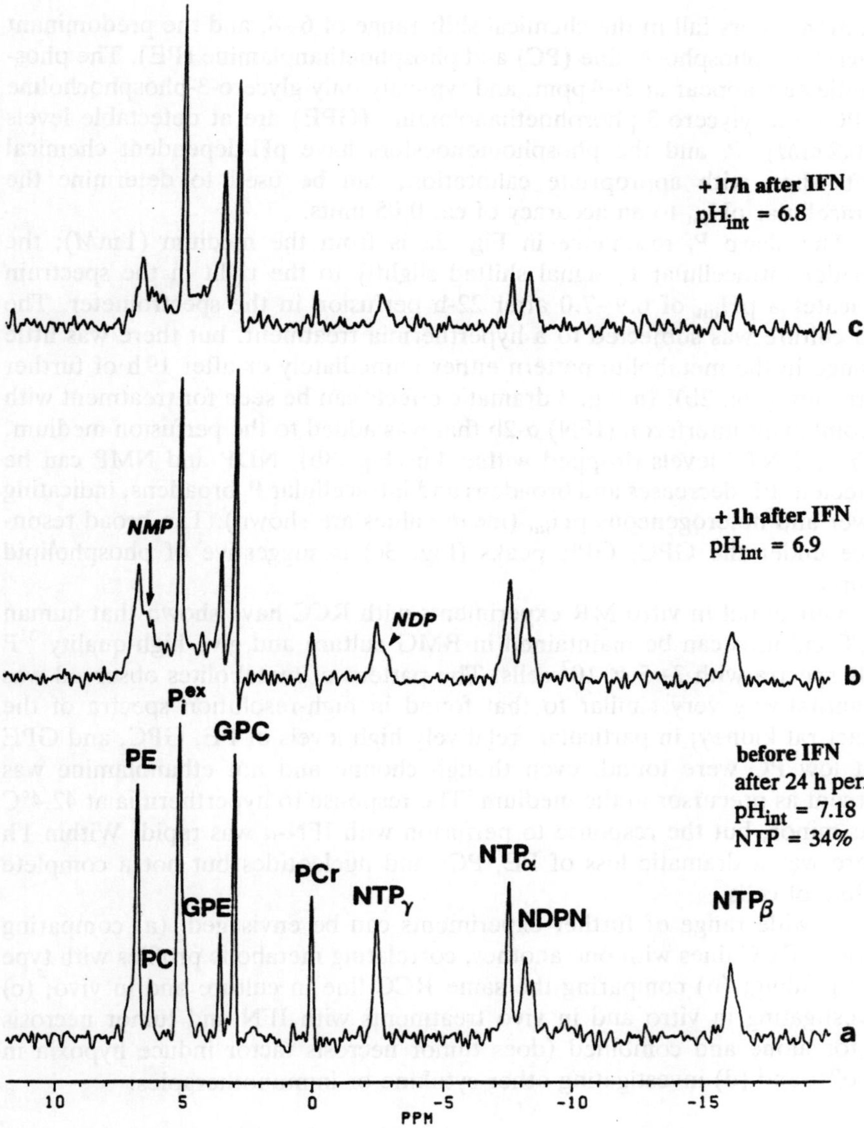

Fig. 3a–c. 202-MHz ^{31}P MR of perfused KTCTL-26A cells as in Fig. 2 but showing the influence of recombinant interferon α-2b (Intron A, Essex Pharma) treatment (5 × 10^6 units/300 ml medium). Cell killing was observed as a significant drop in intracellular pH, a dramatic decrease in NTP and PE levels, and the loss of cells from the gel matrix

phomonoesters fall in the chemical shift range of 6–8, and the predominant species are phosphocholine (PC) and phosphoethanolamine (PE). The phosphodiesters appear at 2–4 ppm, and typically only glycero-3-phosphocholine (GPC) and glycero-3-phosphoethanolamine (GPE) are at detectable levels (>0.2 m*M*). P_i and the phosphomonoesters have pH-dependent chemical shifts that, with appropriate calibration, can be used to determine the intracellular pH_{int} to an accuracy of ca. 0.05 units.

The sharp P_i resonance in Fig. 2a is from the medium (1 m*M*); the broader intracellular P_i signal shifted slightly to the right in the spectrum indicates a pH_{int} of 6.9–7.0 after 22-h perfusion in the spectrometer. The cell culture was subjected to a hyperthermia treatment, but there was little change in the metabolite pattern either immediately or after 19 h of further perfusion (Fig. 2b). In Fig. 3 dramatic effects can be seen for treatment with recombinant interferon (IFN) α-2b that was added to the perfusion medium. PCr and NTP levels dropped within 1 h (Fig. 3b); NDP and NMP can be detected. PE decreases and broadens and intracellular P_i broadens, indicating lower and heterogeneous pH_{int} (mean values are shown). The broad resonance under the GPC, GPE peaks (Fig. 3c) is suggestive of phospholipid debris.

Our initial in vitro MR experiments with RCC have shown that human RCC cultures can be maintained in BMG culture and give high-quality ^{31}P MR spectra with $3–5 \times 10^7$ cells. The pattern of metabolites observed was quantitatively very similar to that found in high-resolution spectra of the intact rat kidney; in particular, relatively high levels of PE, GPC, and GPE but low PC were found, even though choline and not ethanolamine was present as precursor in the medium. The response to hyperthermia at 42.4°C was minor, but the response to perfusion with IFN-α was rapid. Within 1 h there was a dramatic loss of PE, PCr and nucleotides but not a complete killing of cells.

A wide range of further experiments can be envisaged: (a) comparing various RCC lines with one another, correlating metabolic profiles with type and grading; (b) comparing the same RCC line in culture and in vivo; (c) investigating in vitro and in vivo treatments with IFN and tumor necrosis factor alone and combined (does tumor necrosis factor induce hypoxia in vivo?); and (d) investigating other cytokine or immunotherapies.

^{13}C MR Determination of Glycogen in Kidney Tumor Biopsies

Normal kidney tissue has very low levels of stored glycogen (<10 μmol/g), but in renal carcinomas glycogen content can be quite high. Glycogen storage appears to be a general phenomenon associated with the early stages of malignant transformation, and there may be a correlation between glycogen content and histological differentiation or tumor grading.

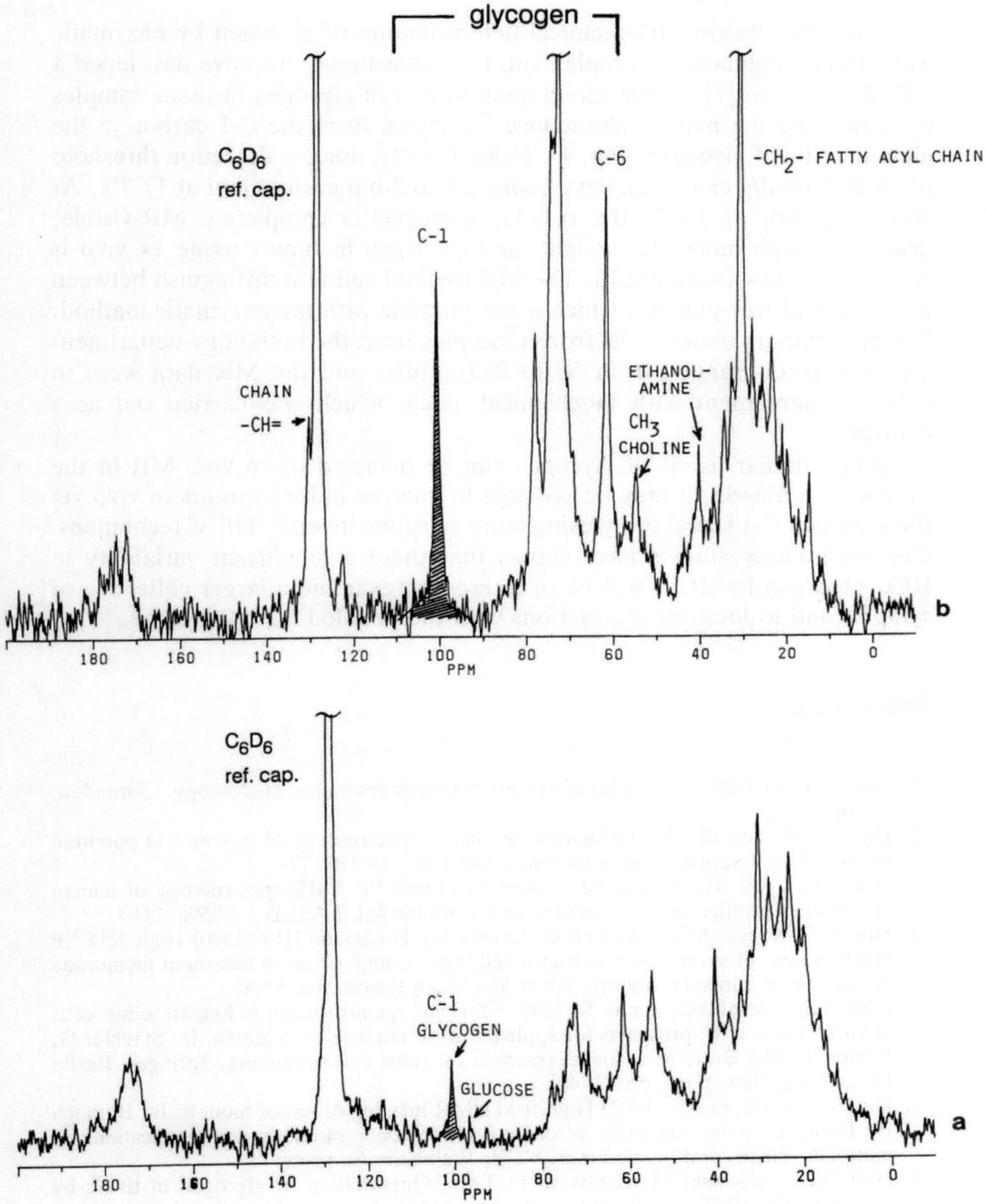

Fig. 4a,b. 125-MHz (11.7-T) natural abundance ^{13}C MR spectra of human renal carcinoma biopsy samples measured in 10-mm MR tubes at 30°C. The C-1 carbon signal from the glucose units in glycogen is shaded. **a** 0.74 g sample of a clear cell carcinoma measured for 2 h, glycogen content = 47 μmol/g, glucose = 10 μmol/g. **b** 0.64 g sample of a clear cell/anaplastic carcinoma measured for 30 min; glycogen content = 330 μmol/g, glucose = 6 μmol/g. A Bruker AM-500 spectrometer was used (inverse-gated ^{1}H decoupling, TR = 1.5 s, 70° flip angle)

Since the classical biochemical determination of glycogen by enzymatic assay of homogenates is complex and time consuming, we have developed a ^{13}C MR method [7] for the direct quantitation of glycogen in tissue samples by measuring the natural abundance ^{13}C signal from the C-1 carbon in the glucose units of glycogen (Fig. 4). Using 0.5–1 g tissue a detection threshold of 10 ± 5 μmol/g can be achieved with a 1 to 2-h measurement at 11.7 T. At 30°C (but not at 10°C) the cellular glycogen is completely MR-visible, despite its high molecular weight, and its decay in tumor tissue ex vivo is remarkably slow (<3 μmol/h). The MR method can also distinguish between glycogen and free glucose, which is not possible with the enzymatic method. For nine tumors obtained as frozen samples from the pathology department glycogen levels ranged from 50 to 350 μmol/g, and the MR data were in excellent agreement with biochemical assay, which was carried out as a control.

Since similar levels of glycogen can be detected by in vivo MR in the liver and in muscle, it may be possible to analyze kidney tumors in vivo via the glycogen C-1 signal or by using more sensitive inverse $^{1}H/^{13}C$ techniques. Our preliminary studies have shown that there is significant variability in RCC glycogen levels. It will be of interest to examine a larger collective of samples and to look for correlations with the detailed histopathology.

References

1. Callaghan PT (1991) Principles of nuclear magnetic resonance microscopy. Clarendon, Oxford
2. Daly PF, Cohen JS (1989) Magnetic resonance spectroscopy of tumors and potential in-vivo clinical applications: a review. Cancer Res 49:770–779
3. Daly PF, Lyon RC, Straka EJ, Cohen JS (1988) ^{31}P NMR spectroscopy of human cancer cells proliferating in a basement membrane gel. FASEB J 2:2596–2604
4. Hull WE, Kuesel AC, Graschew G, Lorenz W, Thielmann HW (1990) High-field ^{31}P NMR studies of several human tumor cell lines. Comparisons in basement membrane gel culture and in solid tumors. Abstr Soc Magn Reson Med 9:845
5. Hull WE, Kuesel AC, Pomer S (1992) ^{31}P NMR spectrosocopy of human tumor cells in vitro and in vivo: prospects for applications to renal cell carcinoma. In: Staehler G, Pomer S (eds) Basic and clinical research on renal cell carcinoma. Springer, Berlin Heidelberg, New York, pp 37–48
6. Hull WE, Fichtner KP (1992) High-field NMR imaging of tumor models. In: Blümich B, Kuhn W (eds) Magnetic resonance microscopy: methods and applications in materials science and biomedicine. VCH, Weinheim (in press)
7. Hull WE, Zerfowski M, Bannasch P (1989) Quantitation of glycogen in tissue by means of ^{13}C-NMR: application to human kidney tumors. Abstr Soc Magn Reson Med 6:488
8. Minichiello MM, Albert DM, Kolodny NH, Lee MS, Craft JL (1989) A perfusion system developed for ^{31}P NMR study of melanoma cells at tissue-like density. Magn Reson Med 10:96–107
9. Narayan KS, Moress EA, Chatham JC, Barker PB (1990) ^{31}P NMR of mammalian cells encapsulated in alginate gels utilizing a new phosphate-free perfusion medium. NMR Biomed 3:23–26

10. Neeman M, Rushkin E, Kadouri A, Degani H (1988) Adaptation of culture methods for NMR studies of anchorage-dependent cells. Magn Reson Med 7:236–242
11. Steen RG (1989) Response of solid tumors to chemotherapy monitored by in-vivo ^{31}P nuclear magnetic resonance spectroscopy: a review. Cancer Res 49:4075–4085

^{31}P Magnetic Resonance Studies of Cytokine-Induced Alterations in Tumor Cells Transplanted into Syngeneic Mice

F. Podo[1], G. Carpinelli[1], E. Proietti[2], and F. Belardelli[2]

Introduction

By monitoring in real time metabolic intermediates and reaction fluxes in intact tissues, localized magnetic resonance spectroscopy (MRS) opens new perspectives for the early detection of tumor response to therapy [6]. ^{31}P MR spectra of neoplastic tissues do not generally exhibit tumor-specific resonances. However, in comparison with healthy tissues, significant alterations can be detected in peak areas of metabolites involved in cell bioenergetics (nucleoside triphosphates, phosphocreatine, inorganic phosphate) or in phospholipid biosynthesis and turnover. In particular, ^{31}P MR spectra of experimental and clinical tumors generally exhibit elevated phosphomonoester (PME) and phosphodiester (PDE) signals, primarily arising from phospholipid precursors and/or catabolites [such as phosphorylcholine (PCho), phosphorylethanolamine (PEtn), glycerophosphorylcholine (GroPCho), glycerophosphorylethanolamine (GroPEtn) and *sn*-glycerol 3-phosphate (GroP)]. PME and PDE peak areas generally exhibit large variability in different tumors and may undergo significant alterations during tumor growth or in response to therapy.

The levels of PCho and PEtn may vary according to cell type and strain, and also depend upon cell growth conditions, such as in vitro or in vivo substrate availability or time of cell harvest from culture medium. At the field strength utilized for clinical studies ($\leq 2\,T$) PCho and PEtn give closely overlapping peaks, while GroPCho and GroPEtn are generally obscured by the broad resonance band of phospholipid headgroups. These resonances can, however, be resolved in experimental tumors, at the higher magnetic fields utilized for experimental studies (e.g., 7–9 T). ^{1}H MRS signals of GroPCho, PCho, and choline are separated in aqueous solution by less than 0.05 ppm and in tissues give rise to only one unresolved resonance, usually referred to as the "choline peak" (3.2 ppm). The area of this resonance may exhibit significant variations from tumor to tumor or in response to therapy. These findings stimulated new interest in the mechanisms underlying MR

[1] Laboratory of Cell Biology and [2] Laboratory of Virology, Istituto Superiore di Sanita', Viale Regina Elena 299, I-00161 Roma, Italy

spectral alterations, in view of a possible use of signals arising from phospholipid metabolites as "fingerprints" of cell proliferation and tumor growth and/or regression.

^{31}P and ^{1}H MRS studies carried out in our laboratories [3, 8–10] on experimental tumors isolated from syngeneic mice at different time intervals after in vivo administration of cytokines [interferon (IFN) α/β, tumor necrosis factor (TNF) α or β, interleukin (IL) 1β] demonstrated that all these antitumor agents induce similar, early (prenecrotic) ^{31}P and ^{1}H spectral modifications. The latter were interpreted on the basis of alterations of some specific enzymatic activities in the pathway of de novo biosynthesis and turnover of phospholipids. The observed MRS alterations were associated with other metabolic and morphological modifications, such as (a) early decrease in ATP, pH alkaline shift, decrease of phospholipids and diacylglycerols (but not triacylglycerols), increase in free fatty acids, and (b) cell disaggregation, vascular congestion, and focal extravasation of erythrocytes.

This paper summarizes the main results obtained in our laboratories from such MR, biochemical, and histological studies. The observed metabolic alterations propose a reexamination of the complex and interlinked biochemical network responsible for the de novo biosynthesis and catabolism of phospholipids, also in the light of recent evidence [7] pointing to a possible role of phospholipid turnover in controlling cell growth and proliferation.

Materials and Methods

Mice. Male or female DBA/2 mice were obtained from Charles River Italia (Milan, Italy). C3H/HeN and C3H/HeJ mice bearing methylcholanthrene-induced fibrosarcoma in the hind leg were kindly provided by Dr. A. Mantovani (Istituto Mario Negri, Milan, Italy).

Tumor Cells. IFN-α/β sensitive (745) or resistant (3Cl-8) Friend erythroleukemia cells (FLC) were passaged serially by intraperitoneal inoculation of DBA/2 mice [1]. These cells have been shown to be resistant in vitro to the cytotoxic effect of TNF. Murine fibrosarcoma cells HeN4 and HeJ16 were kindly provided by Dr. G. Taraboletti and Dr. A. Mantovani (Istituto Mario Negri, Milan). Trypsinized tumor cells were passaged subcutaneously in syngeneic C3H/HeN mice.

Tumor Preparation. FLC solid tumors were obtained in DBA/2 mice by injecting 5 × 10^{6} FLC subcutaneously in the dorsum, as previously described [2]. Fibrosarcoma HeN4 or HeJ16 cells were similarly implanted in C3H/HeN or C3H/HeJ mice, respectively. MR spectra of either tumor cells or freshly dissected tissues or tissue extracts were obtained on a Bruker AM 400 WB spectrometer (9.4 T), as described in previous reports [3, 5, 8, 9].

Interferon. Different preparations of IFN were utilized: (a) partially purified murine IFN-α/β (specific activity 2 × 10^7 U/mg protein) prepared from suspension cultures of mouse sarcoma C243 cells inoculated with Newcastle Disease Virus; (b) highly purified mouse IFN-α/β prepared by sequential affinity chromatography (specific activity 5 × 10^8 U/mg protein). Details on the IFN preparations have been reported elsewhere ([9] and references therein).

Tumor Necrosis Factors. Recombinant murine TNF-α expressed in *Escherichia coli*, purified and tested for endotoxin contamination (specific activity of 8 × 10^7 U/mg protein) was a generous gift of Prof. W. Fiers (Ghent, Belgium). In the experiments, TNF was diluted in buffered physiological solution (NaCl 153 m*M*) containing bovine serum albumin (BSA; 100 μg/ml) and used as described elsewhere [8]. Human recombinant TNF-α (8 × 10^7 U/mg protein) was a kind gift of Drs. W. Fiers and J. Tavernier (Ghent, Belgium). TNF-β, kindly provided by Dr. G. Adolf (Boehringer Ingelheim, Vienna, Austria), had a specific activity of 1–2 × 10^8 U/mg protein and was more than 99% pure.

Interleukin-1. Highly purified, electrophoretically pure human recombinant IL-1β (mature fragment 117–269, expressed in *E. coli*) was obtained from SCLAVO (Siena and Cassina de Pecchi, Italy). The preparation had a specific activity of about 5 × 10^8 half-maximal U/mg protein.

Histology. Histological examinations were performed on tissues fixed in 10% formaldehyde, processed for paraffin embedding, cut in 2- to 3-μm-thick slices and stained with hematoxylin and eosin [3, 8, 10].

Results

Interferon

Previous studies by Belardelli et al. [2] demonstrated that daily in vivo injections of purified mouse IFN-α/β into established subcutaneous FLC tumors (either 745 or 3Cl-8 clone) resulted in a marked inhibition of the growth and necrosis and a regression of most tumors. Similar antitumor effects were observed in mice injected with either IFN-sensitive (745) or IFN-resistant (3Cl-8) cells, suggesting host-mediated mechanisms, resulting in tumor cell degeneration ([2] and references therein). After six or seven daily intratumoral injections of IFN-α/β, histological examinations showed extensive areas of tumor cell necrosis in the absence of obvious host cell infiltrates. Metabolic effects induced by IFN were studied by ^{31}P MR on FLC tumors after only two daily injections of IFN, when histological examinations did not show necrotic areas.

Fig. 1. ^{31}P MR spectra (161.9 MHz, 4°C) of 3Cl-8 Friend erythroleukemia cell tumors treated with 0.2 ml IFN (titer 10^6 U/ml) or control material (BSA, 100 µg/ml). *First number in parenthesis*, the day after tumor cell implantation at which IFN treatment was started; second *number*, the number of daily, intratumoral injections of either IFN or BSA. Peak assignments: *a*, phosphorylethanolamine; *b*, phosphrylcholine; *c*, inorganic phosphate; *d*, glycerophosphorylethanolamine; *e*, glycerophosphorylcholine. *Arrow*, spectral contribution of glycerol 3-phosphate

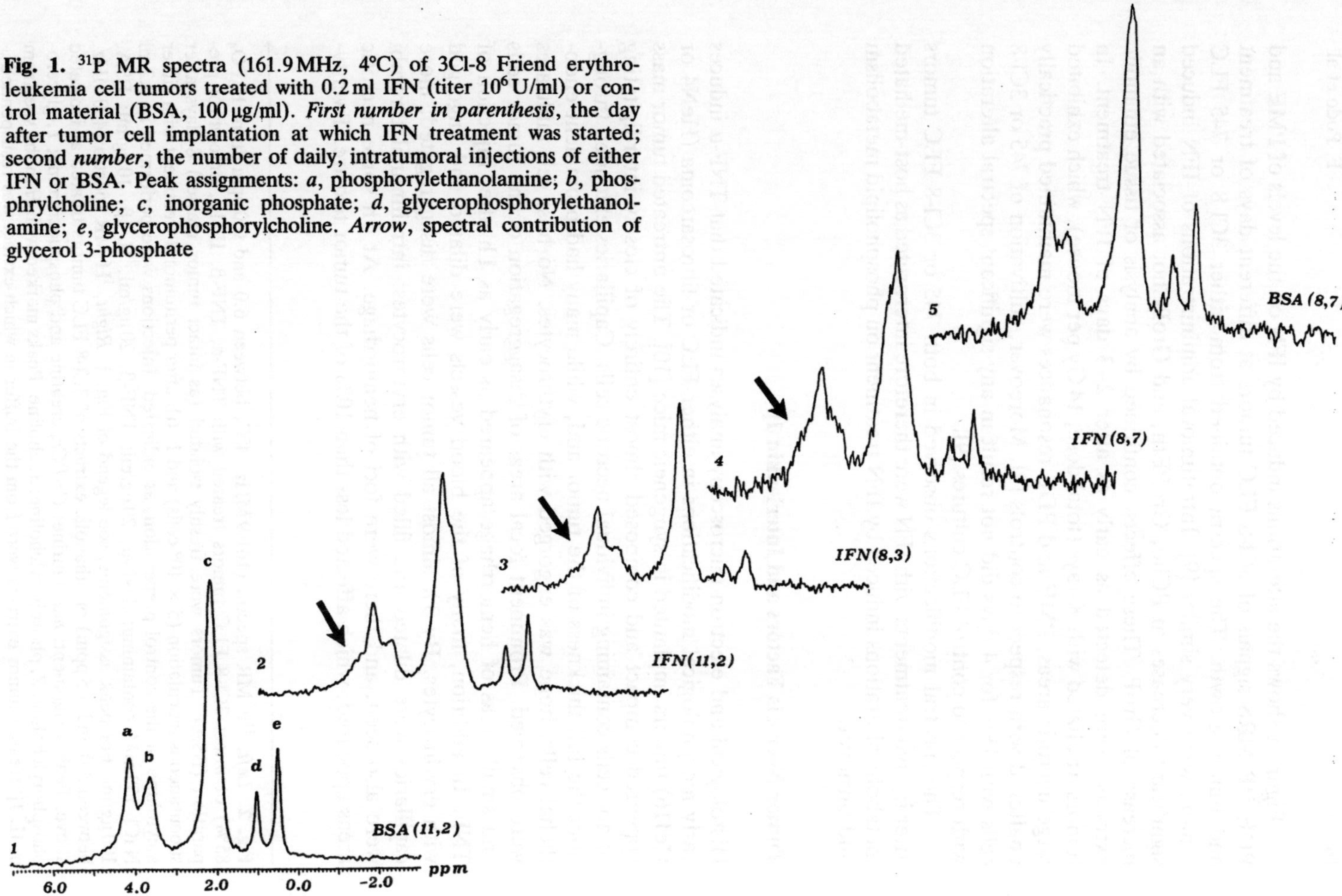

Figure 1 shows the alterations induced by IFN on the levels of PME and PDE ^{31}P MRS signals of 3Cl-8 FLC tumors at different days of treatment and tumor growth. The spectra obtained from either 3Cl-8 or 745 FLC tumors were very similar [9]. Intratumoral administrations of IFN induced significant decreases in PCho, GroPEtn, and GroPCho, associated with an increase of GroP. These effects, confirmed by analysis of tissue ethanolic extracts, were detected as early as after 2–3 days of IFN treatment. In tumors irradiated with X-rays (total dose, 14 Gy per mouse), which exhibited large necrotic areas, PME and PDE resonances were maintained practically unaltered with respect to controls [9]. Moreover, cultivation of 745 or 3Cl-8 cells with IFN for 4 days did not result in any significant spectral alteration with respect to control FLC cultures [9].

The spectral modifications observed in both 745 or 3Cl-8 FLC tumors after in vivo treatments with IFN were therefore interpreted as host-mediated metabolic alterations induced by IFN treatment on phospholipid metabolism and turnover.

Tumor Necrosis Factors and Interleukin 1β

Histological and electron microscopy analyses indicated that TNF-α induces early morphological modifications in either FLC or fibrosarcoma (HeN4 or HeJ16) tumors implanted in syngeneic mice [10]. The untreated tumor mass appeared compact and composed almost entirely of closely interdigitating tumor cells containing individual necrotic cells. Capillaries extended throughout the full thickness of the tumor and, while many had prominent endothelial cells, none was engorged with erythrocytes. No host cell infiltrates were observed. Prominent focal areas of disaggregation of the tumor cells and small areas of hemorrhage appeared as early as 1 h after injection of TNF. In addition, many of the blood vessels were dilated and engorged with erythrocytes. By 3 h almost all tumor cells were disaggregated; more capillaries were dilated and filled with erythrocytes; intraluminal thrombi were also seen, and there were foci of hemorrhage. At 6 h some necrotic areas appeared, which affected less than 10% of the tumor texture. Twenty-

→

Fig. 2. *Left*, ^{31}P MR spectra (161.9 MHz, 4°C, between 6.0 and 0.0 ppm from H_3PO_4 85%) of intact 3Cl-8 FLC tumors treated with TNF-α, TNF-β, IL-1β, or control preparation (BSA). Tumors were freshly isolated (as intact tumor masses) 8 days after subcutaneous inoculation (5×10^6 cells) and 1–6 h after peritumoral treatment with either a cytokine or the control preparation, as indicated. Injections were performed in 0.2 ml NaCl 153 m*M* containing: TNF-α, 20 μg/ml; TNF-β, 20 μg/ml; IL-1β, 10 μg/ml; BSA, 100 μg/ml. For peak assignments, see legend of Fig. 1. *Right*, ^1H MR spectra (400 MHz, between 3.0 and 3.5 ppm) of ethanolic extracts of 3Cl-8 FLC tumors treated as indicated above. Peak assignment: *tau*, taurine; *Cr*/*PCr*, creatine and phosphocreatine; *1*, glycerophosphorylcholine, *2*, phosphorylcholine; *3*, choline. Peaks marked with *H* in the spectrum of IL-1β treated tumor extracts were from the buffer in which extract was resuspended

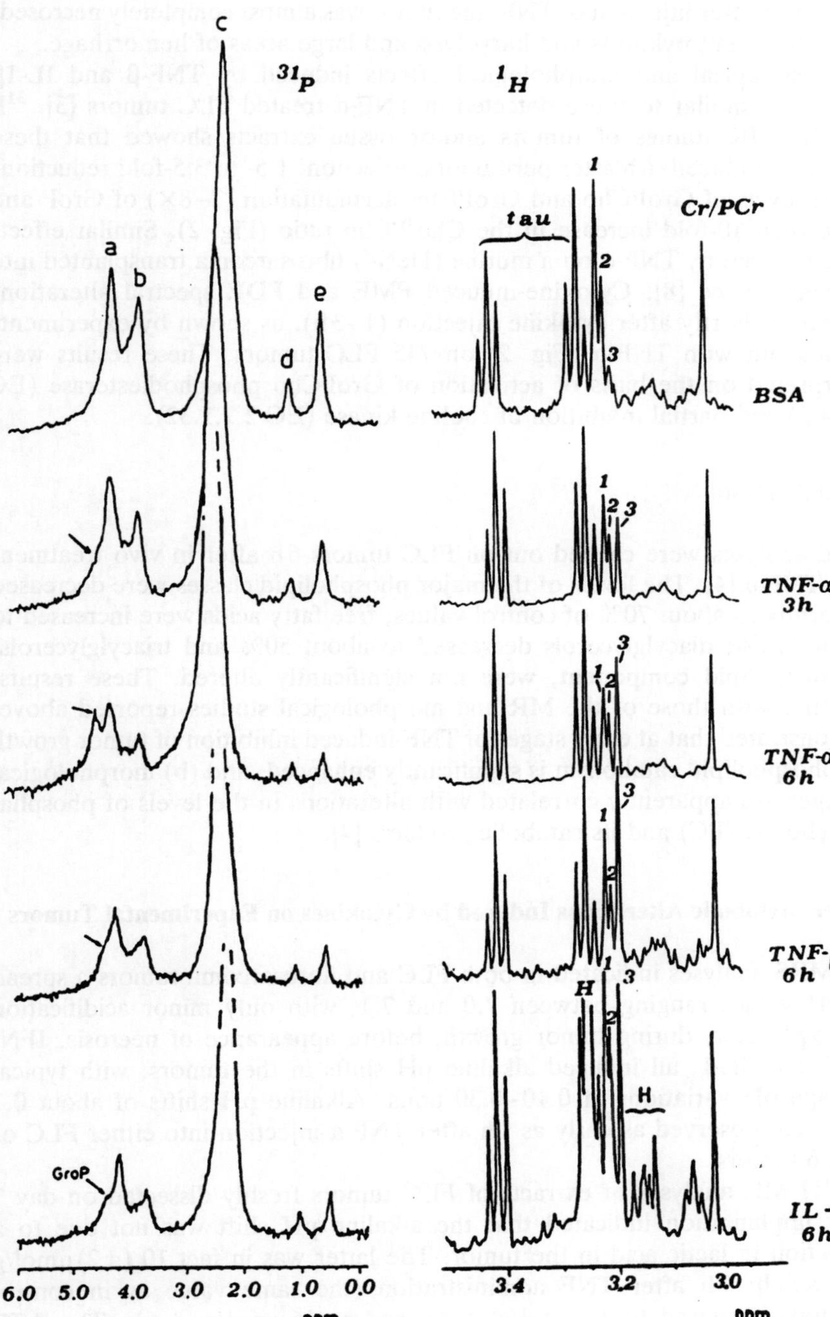

c
31P
1H
a
b
e
d
tau
Cr/PCr
1
2
3
BSA
TNF-α
3h
TNF-α
6h
TNF-β
6h
H
GroP
IL-1β
6h
6.0
5.0
4.0
3.0
2.0
1.0
0.0
ppm
3.4
3.2
3.0
ppm

four hours after injection of TNF, the tumor was almost completely necrosed, with prominent pyknosis and karyolysis and large areas of hemorrhage.

Histological and morphological effects induced by TNF-β and IL-1β were very similar to those detected in TNF-α treated FLC tumors [3]. ^{31}P and ^{1}H MR studies of tumors and/or tissue extracts showed that these cytokines induced, 6 h after peritumoral injection, 1.5- to 3.5-fold reductions in the levels of GroPCho and GroPEtn, accumulation (7–8×) of GroP and more than 10-fold increase in the Cho/PCho ratio (Fig. 2). Similar effects were induced by TNF-α on a murine (HeN4) fibrosarcoma transplanted into syngeneic mice [8]. Cytokine-induced PME and PDE spectral alterations appeared shortly after cytokine injection (1–3 h), as shown by experiments carried out with TNF-α (Fig. 2) on 745 FLC tumors. These results were interpreted on the basis of activation of GroPCho phosphodiesterase (EC 3.1.4.2) and partial inhibition of choline kinase (EC 2.7.1.32).

Lipid Analyses

Lipid analyses were carried out on FLC tumors 6 h after in vivo treatment with TNF-α [4]. The levels of the major phospholipid classes were decreased uniformly to about 70% of control values; free fatty acids were increased to about 170%; diacylglycerols decreased to about 50% and triacylglycerols, the main lipid component, were not significantly altered. These results, together with those of the MR and morphological studies reported above, demonstrated that at early stages of TNF-induced inhibition of tumor growth (a) phospholipid catabolism is significantly enhanced, and (b) morphological changes are apparently correlated with alterations in the levels of phosphatidylcholine (PC) and its catabolic products [4].

Other Metabolic Alterations Induced by Cytokines on Experimental Tumors

^{31}P MRS analyses indicated in both FLC and fibrosarcoma tumors a spread of pH values ranging between 7.0 and 7.1, with only minor acidification (<0.5 pH unit) during tumor growth, before appearance of necrosis. IFN, TNF, and IL-1, all induced alkaline pH shifts in the tumors, with typical average pH variations of 0.10–0.30 units. Alkaline pH shifts of about 0.1 unit were observed as early as 1 h after TNF-α injection into either FLC or HeJ16 tumors.

^{1}H MR analyses of extracts of FLC tumors freshly dissected on day 7 after implantation indicated that the alkaline pH shift was not due to a reduction in lactic acid in the tumor. The latter was in fact 10 (±2) μmol/g wet weight 1 h after TNF administration (the same value as in control tumors), increased to 12 μmol/g at 3 h and to 16 μmol/g at 6 h. The ATP levels decreased by about 70% with respect to control values at 1 h and to 30% ± 15% at 6 h. These findings were consistent with blood vessel occlusion detected in histological sections under similar experimental conditions.

Discussion

The detection in experimental tumors of elevated ^{31}P MRS signals arising from phospholipid precursors and catabolites, and the observation of significant alterations in the pools of these metabolites in cytokine-treated tumors, stimulate new interest in the regulation of phospholipid biosynthesis and turnover in different conditions of cell growth and proliferation.

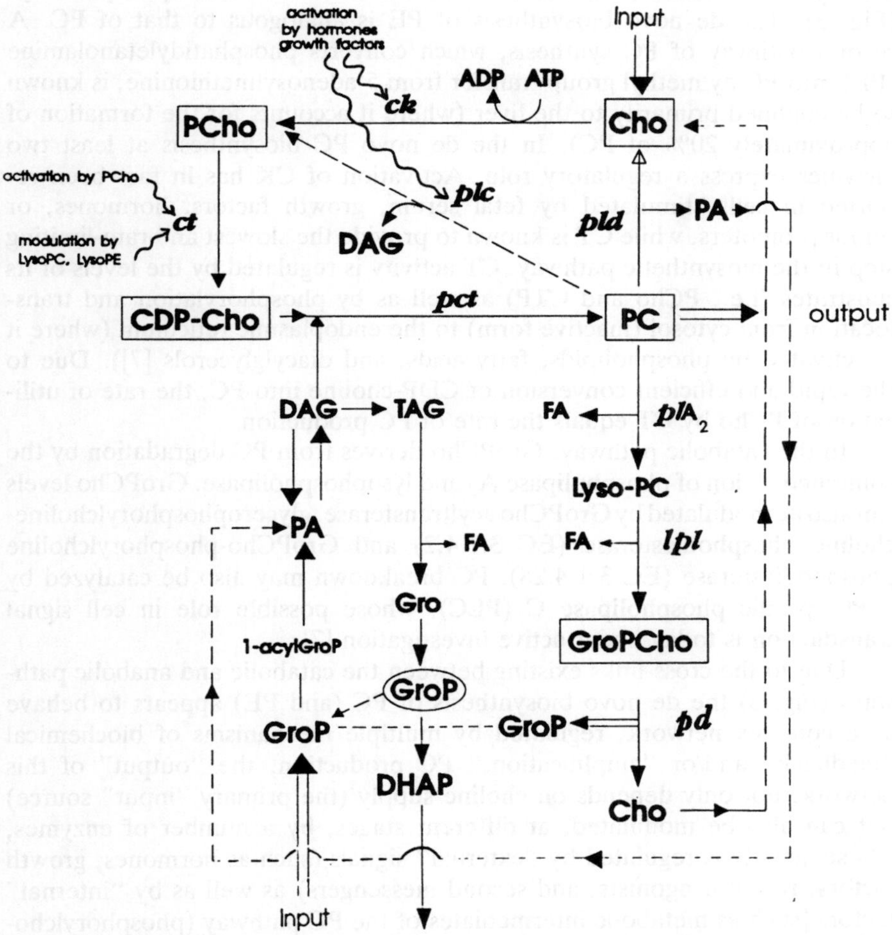

Fig. 3. Biochemical network and regulation of phospholipid biosynthesis and breakdown. Metabolites: *Cho*, choline; *PCho*, phosphorylcholine; *CDP-Cho*, cytidine diphosphate choline; *PC*, phosphatidylcholine; *Lyso-PC*, lysophosphatidylcholine; *GroPCho*, glycerophosphorylcholine; *GroP*, glycerol 3-phosphate; *PA*, phosphatidate; *DAG*, diacylglycerol; *TAG*, triacylglycerol; *Gro*, glycerol; *FA*, fatty acids; *DHAP*, dihydroxyacetone phosphate. Enzymes: *ck*, choline kinase; *ct*, cytidylyltrasferase; *pct*, phosphocholine transferase; *plc*, phospholipase C; *pld*, phospholipase D; *pla*, phospholipase A; *lpl*, lysophosphatidylcholine phospholipase; *pd*, glycerophosphorylcholine choline phosphodiesterase

According to classical biochemistry, the major pathway of PC biosynthesis is classically described by a three-step pathway, regulated by the activity of three enzymes: (a) ATP: choline phosphotransferase (EC 2.7.1.32) or choline kinase (CK), responsible for the phosphorylation of choline into PCho; (b) CTP: choline phosphate cytidylyltransferase (CT), producing cytidine disphosphate choline (CDP-choline) from CTP and PCho (EC 2.7.7.15); (c) CDP-choline: 1,2-diacylglycerol choline phosphotransferase (EC 2.7.8.2) or phosphocholine transferase. The latter enzyme expresses its activity at the branch point between the PC and the neutral lipids pathways (Fig. 3). The de novo biosynthesis of PE is analogous to that of PC. A second pathway of PC synthesis, which converts phosphatidyletanolamine (PE) into PC by methyl group transfer from 5-adenosylmethionine, is known to be confined primarily to the liver (where it accounts for the formation of approximately 20% of PC). In the de novo PC biosynthesis at least two enzymes express a regulatory role. Activation of CK has in fact been reported in cells stimulated by fetal serum, growth factors, hormones, or tumor promoters, while CT is known to provide the slowest and rate-limiting step in the biosynthetic pathway. CT activity is regulated by the levels of its substrates (i.e., PCho and CTP) as well as by phosphorylation and translocation from cytosol (inactive form) to the endoplasmic reticulum (where it is activated by phospholipids, fatty acids, and diacylglycerols [7]). Due to the rapid and efficient conversion of CDP-choline into PC, the rate of utilization of PCho by CT equals the rate of PC production.

In the catabolic pathway, GroPCho derives from PC degradation by the combined action of phospholipase A_2 and lysophospholipase. GroPCho levels can also be modulated by GroPCho acyltransferase, glycerophosphorylcholine-choline phosphodiesterase (EC 3.1.4.2) and GroPCho-phosphorylcholine phosphodiesterase (EC 3.1.4.28). PC breakdown may also be catalyzed by a PC-specific phospholipase C (PLC), whose possible role in cell signal transduction is today under active investigation [7].

Due to the cross-links existing between the catabolic and anabolic pathways (Fig. 3) the de novo biosynthesis of PC (and PE) appears to behave as a complex network, regulated by multiple mechanisms of biochemical "feedback" and/or "amplification." PC production, the "output" of this network, not only depends on choline supply (the primary "input" source) but can also be modulated, at different stages, by a number of enzymes, whose activity is regulated by "external" signals (such as hormones, growth factors, receptor agonists, and second messengers) as well as by "internal" factors [such as metabolic intermediates of the PC pathway (phosphorylcholine, lysophospholipis, fatty acids, diacylglycerols), substrate supply, ATP, acetyl-coenzyme A, etc.). Some of the existing "cross-links" between the anabolic and catabolic PC pathways are indicated in Fig. 3. On the basis of these considerations, similarly to an electronic network, the PC pathway seems capable of undergoing transitions from a "steady-state" condition (normal phospholipid turnover and cell growth) to a "high-gain amplifier" regimen, in which enhanced PC production and turnover contributes to fast cell proliferation (as in tumor or embryonic tissues). In this condition, high levels of PCho and/or PEtn (and therefore elevated ^{31}P MRS PME resonances) might be seen as "indicators" of accelerated PC production and turnover. Not only, in fact, do high PCho levels contribute to stimulate CT activity, but they may also be the result of stimulation of CK and/or PLC. On the other hand, in the phosphodiester spectral region GroPCho and GroPEtn signal areas may provide interesting "markers" of cell maturation and/or inhibition of tumor growth. The levels of these PDE compounds have in fact been reported to undergo significant reductions following in vitro induced cell differentiation [5], as well as upon cell transition from highly proliferative to more mature stages.

In all cytokine-treated tumors the levels of GroPCho were found to be significantly reduced, with concomitant increase in the concentration of the two hydrolytic products choline and glycerol 3-phosphate. These results pointed to cytokine-induced activation of GroPCho phosphodiesterase, an enzyme so far detected and studied only in mature, normal tissues. Recent studies in our laboratories demonstrated that this enzyme is also activated in FLC upon in vitro induced erythroid differentiation [11]. In differentiated cells the activity of this enzyme was very close to that of normal cells (e.g., hepatocytes), whereas in undifferentiated cells it was about three times lower. It can be speculated that a low activity of GroPCho diesterase (and therefore a limited PC breakdown along the classical catabolic pathway) might be "beneficial" to a fast growing cell, by limiting PC loss and/or by facilitating preferential PC breakdown by the alternative, PLC-regulated, catabolic pathway. The activation of GroPCho diesterase in cytokine-treated tumors would also support the attractive hypothesis that these cytokines may contribute to induce, in malignant cells, a transition toward more mature, less proliferative stages.

Acknowledgement. This work was carried out with partial support by the CNR Special Project on Oncology (1984–1988), the Associazione Italiana per la Ricerca sul Cancro, the Italy-USA Project Therapy of Tumors and the Comitato Italiano per le Scienze Biologiche e Mediche CNR (grant no. 89.04697.CT04). We thank Dr. M. Leach for discussions on phospholipid metabolism in tumors, and Mr. M. Giannini for high-level technical assistance.

References

1. Belardelli F, Ferrantini M, Maury C, Santurbano L, Gresser I (1984) Biological and biochemical differences between in vitro and in vivo passaged Friend leukemia cells: I. Tumorigenicity and capacity to metastasize. Int J Cancer 34:389–395

2. Belardelli F, Gresser I, Maury C, Maunoury MT (1983) Antitumor effects of interferon in mice injected with interferon-sensitive and interferon-resistant Friend leukemia cells: III. Inhibition of growth and necrosis of tumours implanted subcutaneously. Int J Cancer 31:649–653
3. Belardelli F, Proietti E, Ciolli V, Sestili P, Carpinelli G, Di Vito M, Ferretti A, Woodrow D, Boraschi D, Podo F (1989) Interleukin-1β induces tumor necrosis and early morphologic and metabolic changes in transplantable mouse tumors. Similarities with the anti-tumor effects of tumor necrosis factor α or β. Int J Cancer 44:116–123
4. Bravo E, Carpinelli G, Proietti E, Belardelli F, Cantafora A, Podo F (1990) Alterations of lipid composition in Friend leukemia cell tumors in mice treated with tumor necrosis factor-α. FEBS Lett 260:220–224
5. Carpinelli G, Podo F, Di Vito M, Proietti E, Gessani S, Belardelli F (1984) Modulations of glycerophosphorylcholine and phosphorylcholine in Friend erythroleukemia cells upon in vitro-induced erythroid differentiation: a ^{31}P NMR study. FEBS Lett 176:88–92
6. Leach M, Le Moyec L, Podo F (1992) MRS of tumours: basic principles. In: de Certaines JD, Bovee WMMJ, Podo F (eds) Magnetic resonance spectroscopy in biology and medicine. Functional and pathological tissue characterization. Pergamon, Oxford, Chap 18, pp 295–344
7. Pelech SL, Vance DE (1989) Signal transduction via phosphatidylcholine cycles. Trends Biochem Sci 14:28–30
8. Podo F, Carpinelli G, Di Vito M, Giannini M, Proietti E, Fiers W, Gresser I, Belardelli F (1987) Nuclear magnetic resonance analysis of tumor necrosis factor-induced alterations of phospholipid metabolites and pH in Friend leukemia cell tumors and fibrosarcomas in mice. Cancer Res 47:6481–6489
9. Proietti E, Carpinelli G, Di Vito M, Belardelli F, Gresser I, Podo F (1986) ^{31}P-Nuclear magnetic resonance analysis of interferon-induced alterations of phospholipid metabolites in interferon-sensitive and interferon-resistant Friend leukemia cell tumors in mice. Cancer Res 46:2849–2857
10. Proietti E, Belardelli F, Carpinelli G, Di Vito M, Woodrow D, Moss J, Sestili P, Fiers W, Gresser I, Podo F (1988) Tumor necrosis factor α induces early morphologic and metabolic alterations in Friend leukemia cell tumors and fibrosarcomas in mice. Int J Cancer 42:582–591
11. Podo F, Carpinelli G, Ferretti A, Borghi P, Proietti E, Belardelli F (1992) Activation of glycerophosphorylcholine diesterase in Friend leukemia cells upon in-vitro induced eryhtroid differentiation. ^{31}P and ^{1}H NMR studies. Isr J Chem 32:291–298

^{13}C and ^{1}H Magnetic Resonance Studies of Normal and Neoplastic Brain Cells

D. Leibfritz, A. Brand, and C. Richter-Landsberg[1]

Introduction

Current volume-selective in vivo magnetic resonance (MR) spectroscopy in humans records signals from volumes larger than $1\,cm^3$. In this case, the spectra contain peaks from various cell types, which is particularly true for organs such as the brain and kidney. Cell cultures are a means of gaining access to the metabolism of individual cells. As MR requires a fairly large number of cells (approximately 10^8) two strategies may be employed: either cells are grown in several dishes under optimal conditions and then harvested and extracted, or, if enough cells can be supplied adequately in a MR tube, their metabolism is followed in vivo. In the latter case they are grown on microspheres [1], agarose gel [2], or basement membrane gel [3]. Cell extracts have the advantage of less demanding cultural conditions. One may concentrate the extracts from culture dishes with low cell densities. Extract spectra also have a much higher spectral resolution and better signal-to-noise ratio.

Brain tissue consists of two major cell types: the nerve cells or neurons, with numerous afferent and efferent extensions forming the neuritic network, and the glia cells, which do not form synaptic contacts but possess the ability to divide throughout life. The latter not only sustain the neuronal network but play important roles in nerve cell functions and in controlling the neuronal environment. The nervous system thus represents a functional unit based on the dynamic interactions between neurons and glia cells.

Cell cultures represent a useful method for investigating metabolic events in the brain. MR studies can give partial access either to the metabolism of uniform cells or to mixed populations of cells. Although it is possible to prepare primary neuronal cultures of a single phenotype, it is difficult to suppress the growth of unwanted cell populations and rather expensive to yield sufficient amounts of cells. An alternative method is to use clonal cell lines of tumoral origin, with distinct neuronal or glial properties, which divide continuously and have retained their ability under certain growth

[1] University of Bremen, Fachbereich 2, NW2, D-28359 Bremen 33, FRG

conditions to morphologically and biochemically differentiate to their non-neoplastic phenotype and to reduce cell proliferation. These cell lines provide sufficient material for neurochemical studies using MR. Additionally, the comparison of primary cell cultures with cultures of clonal cell lines may indicate tumor-specific metabolic behavior.

The MR studies reported here were carried out using primary cultures of neurons and glia derived from the embryonic or newborn rat brain, and two clonal cell lines, i.e., neuroblastoma N1E-115 and C6 glioma cells. C6 glioma cells were derived from the rat and represent a model system for the investigation of glia cells [4]. N1E-115 is a subclone of a spontaneously developed neuroblastoma in the mouse and is used as a model system for neuronal cells [5]. In order to study their steady-state metabolism, cells were fed with a ^{13}C-labeled substrate such as [1-^{13}C]glucose for a definite period of time. After shock freezing with liquid nitrogen the cells were extracted. ^{13}C and ^{1}H spectra give snapshots of the cell metabolism, and ^{31}P spectra indicate the energy status of the cells at the intervention point.

Materials and Methods

Cell Culture. All cell types described are adherently growing cells, and were kept in plastic culture dishes. The C6 glioma medium consisted of Dulbecco's modified essential medium (DMEM) with 5% fetal calf serum (FCS), whereas the neuroblastoma N1E-115 were grown in DMEM with 10% FCS. Fully differentiated N1E-115 cells were obtained if the FCS concentration was reduced to 0.6% by half dilution with pure DMEM in five steps within 6 days. The morphology changed over this period from a somewhat spherical shape initially to flat cells with the neurites characteristic of neurons. Primary glia cells prepared from 2-day-old newborn rats were incubated in DMEM with 10% FCS for 28 days. Neuronal cultures prepared from 16-day-old rat embryos were grown in BME supplemented with 0.5% FCS for 6 days on poly-L-lysine-coated culture dishes. At this time the cultures contained mainly neurons and only small amounts of glia cells (<10%) [6].

PCA Extract. Approximately 10^8 cells were incubated with fresh medium, free of serum but with 5 mmol [1-^{13}C]glucose. Cultures were washed with ice-cold phosphate-buffered saline (PBS) and immediately frozen with liquid nitrogen. After the addition of cold perchloric acid (12%) the extract was centrifuged, neutralized with KOH, and lyophilized.

Magnetic Resonance. MR spectra were obtained on a Bruker AM-360 from 800 accumulations for ^{31}P (with methylenediphosphonate as reference), 200 accumulations for ^{1}H (with Na-2,2-dimethyl-2-silapentane-5-sulfonate as reference), and 10 000 accumulations for ^{13}C (with dioxane as reference).

Results and Discussion

The ^{31}P spectra of C6 glioma and N1E-115 indicated the normal energy status before the extraction, which means a high concentration of ATP and low amounts of ADP. The ATP/ADP ratio of better than 8:1 indicated a high cell viability. Remarkably, the N1E-115 contained no MR detectable phosphocreatine (PCr; Fig. 1). This was confirmed by the proton spectrum

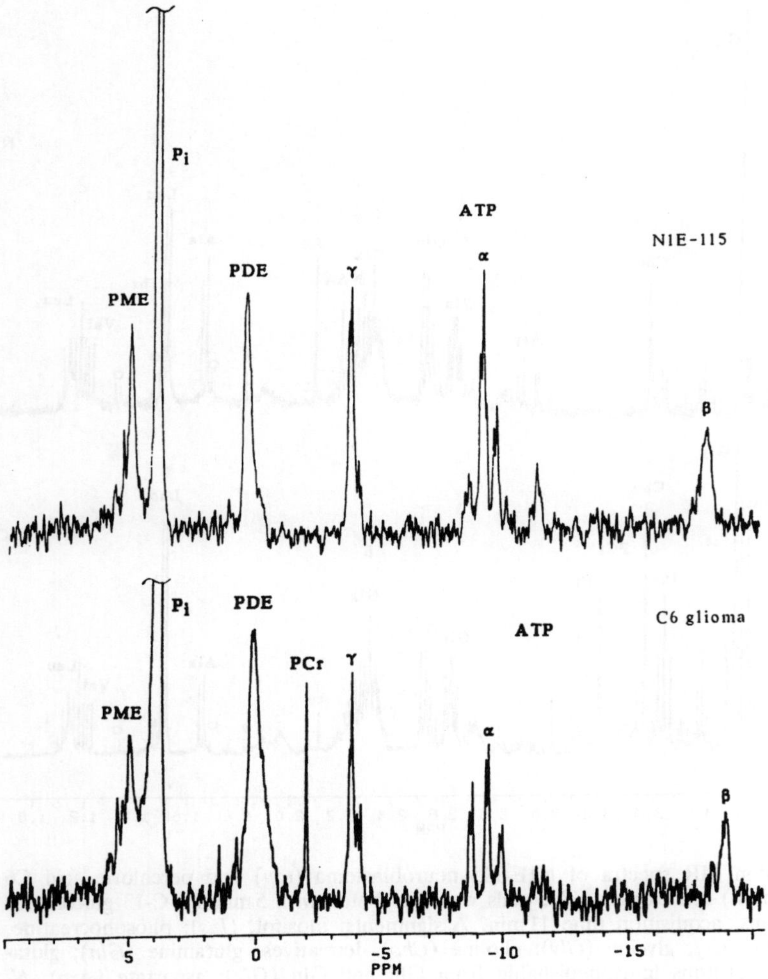

Fig. 1. ^{31}P MR studies from perchloric acid cell extracts of N1E-115 neuroblastoma (*top*) and C6 glioma (*bottom*); 800 scans, spectral acquisition time 46 min. The peaks are assigned as phosphomonoester (*PME*); inorganic phosphate (*Pi*), mainly from the phosphate buffer; phosphodiester (*PDE*); phosphocreatine (*PCr*); adenosintriphosphate (α, β, γ ATP)

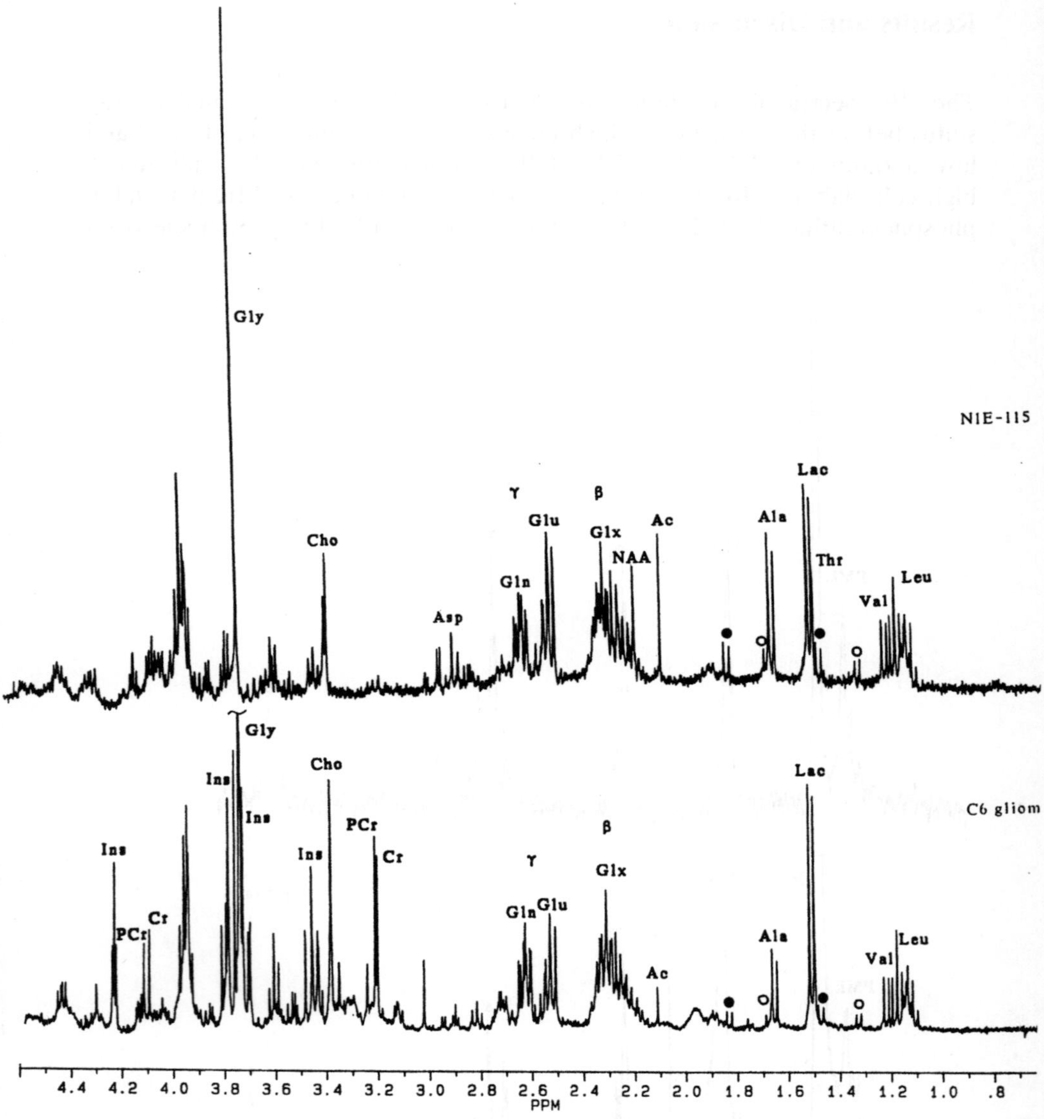

Fig. 2. Proton MR spectra of N1E-115 neuroblastoma (*top*) and perchloric acid C6 glioma (*bottom*) cell extracts. The cells were incubated with 5 mmol 1-C-13 glucose for 2 h; 200 scans, acquisition time 11 min. Assignments: inositol, (*Ins*); phosphocreatine, creatine, (*PCr/Cr*); glycine (*Gly*); choline (*Cho*) derivatives; glutamine (*Gln*); glutamate (*Glu*); protons indistinguishable from Glu and Gln (*Glx*); aspartate (*Asp*); *N*-acetylaspartate (*NAA*); acetate (*Ac*); alanine (*Ala*); doublett splitting of Ala due to ^{13}C enrichment (●); lactate (*Lac*); doublett splitting of Lac due to ^{13}C enrichment (*o*) threonine (*Thr*); valine (*Val*); leucine (*Leu*)

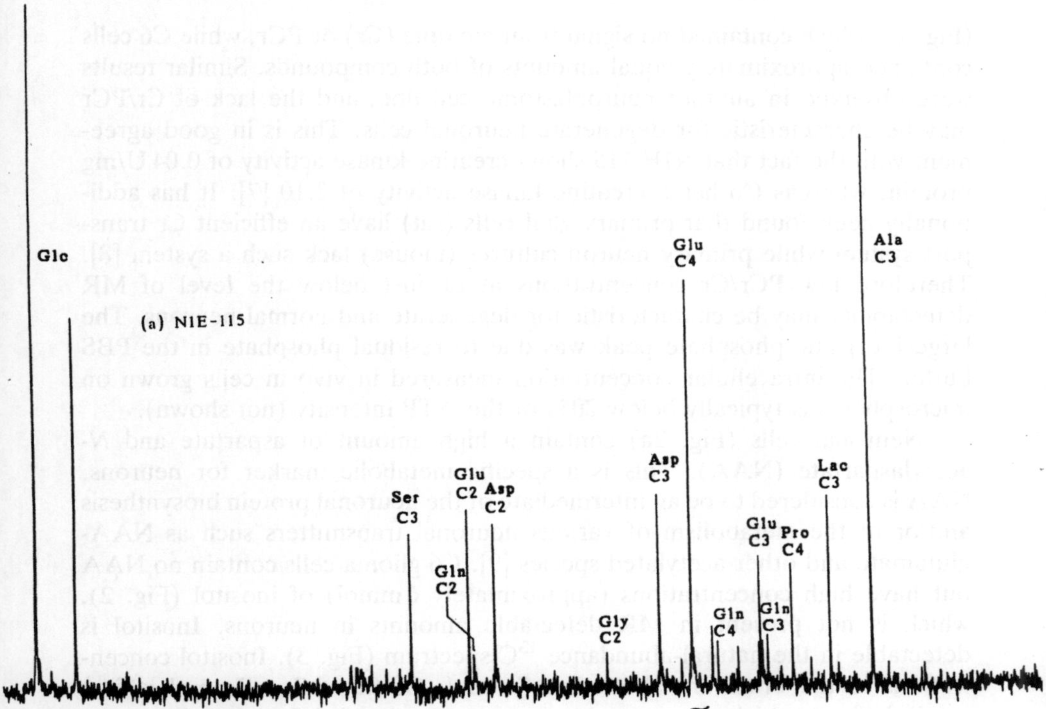

Fig. 3. ^{13}C MR spectra of N1E-115 neuroblastoma (*top*) and perchloric acid C6 glioma (*bottom*) cell extracts. The cells were incubated with 5 mmol 1-C-13 glucose for 2 h; 10 000 scans, acquisition time 6 h 50 min. Assignments: α- and β-anomeric glucose (*Glc*); serine (*Ser*); aspartate (*Asp*); glutamate (*Glu*); glutamine (*Gln*); glycine (*Gly*); lactate (*Lac*); alanine (*Ala*); proline (*Pro*); glycerol-3-phosphate (*Glyc3P*)

(Fig. 2), which contained no signal from creatine (Cr) or PCr, while C6 cells contained approximately equal amounts of both compounds. Similar results were observed in another neuroblastoma cell line, and the lack of Cr/PCr may be characteristic for degenerate neuronal cells. This is in good agreement with the fact that N1E-115 shows creatine kinase activity of 0.04 U/mg protein, whereas C6 has a creatine kinase activity of 2.10 [7]. It has additionally been found that primary glial cells (rat) have an efficient Cr transport system while primary neuron cultures (mouse) lack such a system [8]. Therefore low PCr/Cr concentrations at or just below the level of MR detectability may be characteristic for degenerate and normal neurons. The large inorganic phosphate peak was due to residual phosphate in the PBS buffer. The intracellular concentration measured in vivo in cells grown on microspheres is typically below 20% of the ATP intensity (not shown).

Neuronal cells (Fig. 2a) contain a high amount of aspartate and *N*-acetylaspartate (NAA). This is a specific metabolic marker for neurons. NAA is considered to be an intermediate of the neuronal protein biosynthesis and/or of the metabolism of various neuronal transmitters such as NAA-glutamate and other acetylated species [9]. C6 glioma cells contain no NAA but have high concentrations (approximately 6 mmol) of inositol (Fig. 2), which is not present in MR detectable amounts in neurons. Inositol is detectable in the natural-abundance ^{13}C spectrum (Fig. 3). Inositol concentrations in the brain are about 7–8.7 mmol. Of this 5.2–6.7 mmol is free inositol, from which the second messenger inositol-1,4,5-triphosphate, the phospholipid phosphatidylinositol, and others are formed [10].

The proton and carbon spectra of C6 and N1E-115 also showed differing glutamine concentrations (Fig. 2). Whereas the concentrations of glutamine and glutamate were equal in C6 glioma cells 2 h after the medium exchange, the ratio was 0.5:1 for N115 neuroblastoma cells and dropped even further when the medium was not changed for 24 h (not shown). In brain tissue glia cells regulate the concentration of extracellular glutamate which is released by neurons during neurotransmission. To keep the extracellular glutamate concentration below 10 µmol the glia cells require both an efficient transport system for glutamate and glutamine synthetase to transform it into glutamine. Glutamine is exported again and serves as a substrate for neurons. The extracellular glutamine concentration is 500 µmol [11]. The glutamine synthetase reaction is also part of the ammonic regulation in the brain.

C6 glioma cells contain the typically high lactate concentration of tumor cells, as is seen in the proton and carbon spectra (Figs. 2, 3). Because of their high glycolysis rate tumor cells reduce more pyruvate to lactate than normal cells in order to replenish the NAD^+ pool (Warburg effect). Therefore if FCS is withdrawn from the medium, N1E-115 cells not only change their morphology but also reduce their lactate production.

The fate of the pyruvate formed from 1-^{13}C-labeled glucose during glycolysis can be followed from the spread of the ^{13}C label. Because two pyruvates are generated from one glucose, only half of the pyruvate is marked at the methyl carbon. Pyruvate is transformed in consecutive metabolic steps either to lactate, by lactate dehydrogenase, or to alanine, by alanine transaminase (one of the links between carbohydrate and amino acid metabolism), or most commonly it is introduced into the tricarboxilic acid (TCA) cycle via pyruvate dehydrogenase. The carbon label leaves the TCA cycle by the transamination of α-ketoglutarate to glutamate labeled in position C4 (Fig. 3). The low intensity doublet signal in addition to the glutamate C4 singlet signal with the coupling constant of J = 36 Hz confirms the presence of a C4 and C3 double-labeled glutamate. Therefore part of the label remains in the TCA cycle and forms in a second circulation double-labeled glutamate with lower intensity, because only half of the original pyruvate was labeled [12]. Monolabeled glutamate at C2 can be formed if C4 labeled α-ketoglutarate remains in the TCA cycle for a second circulation and incorporates an unlabeled acetyl-coenzyme A, or alternatively if labeled pyruvate generates oxaloacetate by pyruvate carboxylase activity. The following reactions of the TCA cycle lead to C2-labeled glutamate. If one assumes a fast equilibrium for the malate dehydrogenase and fumarase reactions, then C2 labeled oxaloacetate will also be present in the TCA cycle and finally forms C3 labeled glutamate. It can be seen from the same arguments that C2- or C3-labeled aspartate stems from the corresponding oxaloacetate precursor.

C3-labeled serine is formed from 3-phosphoglycerate via phosphoserine, whereas glycine produced from serine is unlabeled because the serine C3 carbon is split off in a C1 transfer reaction. Proline biosynthesis from glutamate can also be monitored as the C4 carbon of glutamate is transferred to the C4 of proline (see Fig. 3).

The MR spectra of primary glia and neuronal cells (Fig. 4) corresponded well to the differentiated C6 glioma and N1E-115 neuroblastoma spectra (Fig. 3). Primary neurons contained comparable amounts of NAA, glutamine, and lactate (Fig. 4a) while the alanine and aspartate concentrations were lower. In contrast to N1E-115, primary neurons contained a small amount of Cr/PCr.

The content of PCr/Cr in primary neuronal cultures may be due to small amounts of glia cells present in the system or may indeed reflect the presence of minor concentrations of PCr/Cr in neurons. It may be noted that intact brain tissue contains about a three times higher Cr/PCr concentration, which seems to be located primarly in glia cells.

While C3-labeled serine was hardly seen in primary neuronal cells, the C1 signals of both anomeric forms of glucose were exceptionally intense with respect to all other labeled metablites (not shown). Therefore primary neurons had a higher intracellular glucose concentration in comparison to N1E-115 cells, and as expected the metabolic rate was much slower with respect to the tumor line.

While substrate turnover and the labeling experiment with primary neurons corresponded quite well to the behavior of the N1E-115 tumor cell

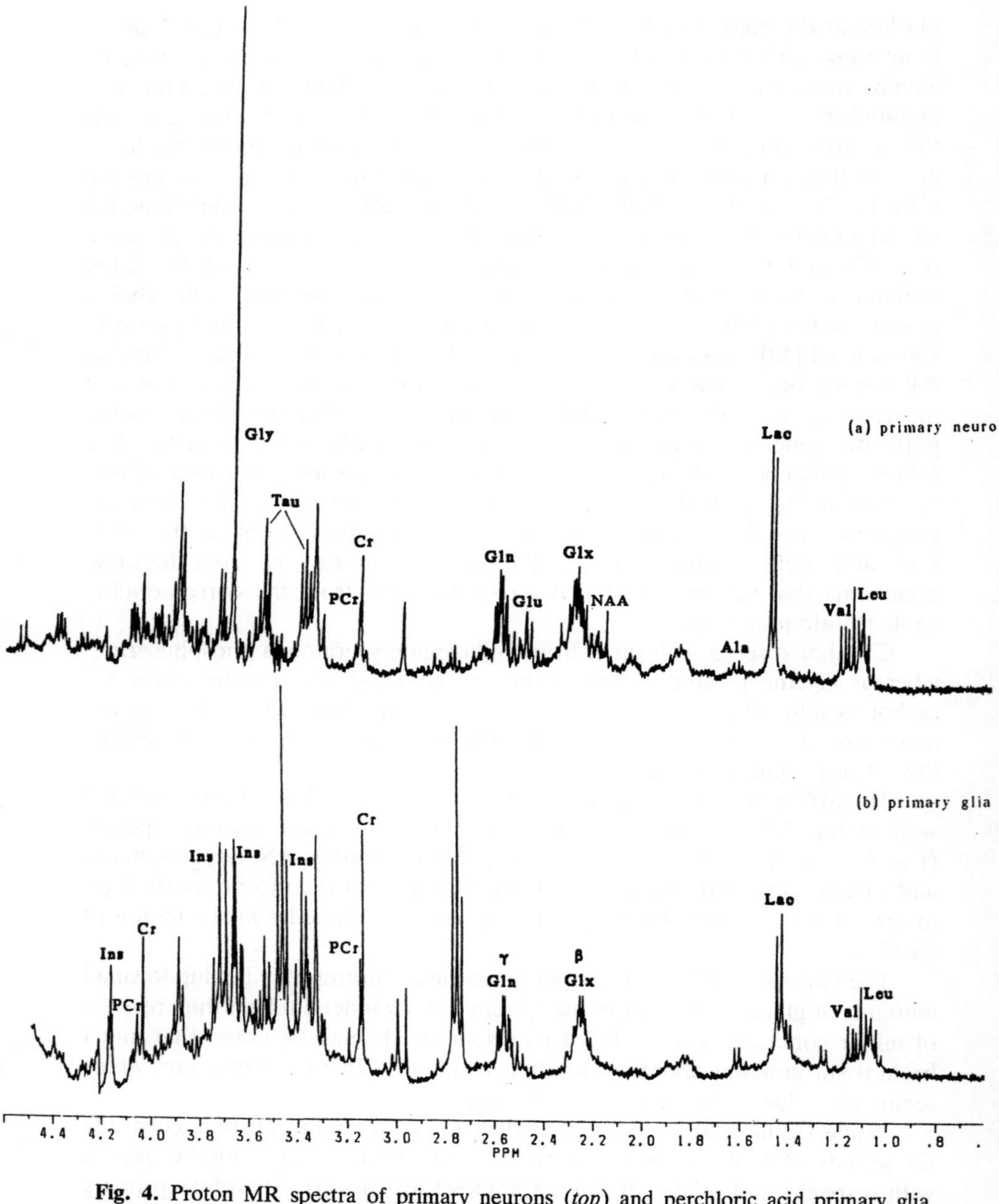

Fig. 4. Proton MR spectra of primary neurons (*top*) and perchloric acid primary glia (*bottom*) cell extracts after incubation with 5 mmol 1-C-13 glucose for 3 h; 200 scans, acquisition time 11 min. Assignment as in Fig. 2; taurine (*Tau*)

line, the metabolic rate of primary glia cells was much slower. Besides labeled glucose the ^{13}C spectrum (not shown) contained only small amounts of lactate and glutamate insetted at C4. Other detectable metabolites had ^{13}C in natural abundance. The proton spectra indicated no alanine, very little glutamate, and twice the amount of Cr/PCr compared to C6 glioma.

In conclusion, MR spectroscopy of cell extracts can be used to compare the bulk metabolism of normal and pathological cells. ^{13}C-labeled precursor molecules were followed through their metabolic pathways, which is not possible in steady-state reactions without isotopic labeling. In contrast to the ^{14}C technique MR avoids radioactive labels and needs no separation techniques such as chromatography. The labeled substrates are detectable in complex mixtures and identified by their chemical shifts. Fluxes and transport phenomenona can be studied, and the energy charge of cells can be determined.

References

1. Fantini J, Galons JP, Marvaldi J, Cozzone PJ, Canioni P (1987) Growth of a human colonic adenocarcinoma cell line (HT29) on microcarrier beads: metabolic studies by ^{31}phosphorus nuclear magnetic resonance spectroscopy. Int J Cancer 39:225–260
2. Cohen JS, Lyon RC, Chen C, Faustino PJ, Batist G, Shoemaker M, Rubalcaba E, Cowan KH (1986) Differences in phosphate metabolite levels in drug-sensitive and -resistant human breast cancer cell lines determined by ^{31}P magnetic resonance spectroscopy. Cancer Res 46:4087–4090
3. Daly PF, Lyon R, Straka EJ, Cohen JS (1988) ^{31}P-NMR spectroscopy of human cancer cells proliferating in a basement membrane gel. FASEB J 2:2596–2604
4. Benda P, Lightbody J, Sato G, Levine L, Sweet W, (1968) Differentiated rat glial cell strain in tissue culture. Science 161:370
5. Nelson PG, Lieberman M (1981) Exitable cells in tissue culture. Plenum, New York, pp 173–245
6. Richter-Landsberg C (1988) NILE glycoprotein in developing rat cerebral cells in culture. Cell Tissue Res 252:181–190
7. Möller A (1989) Untersuchungen über das Enzym Creatinkinase und sein Substrat Creatin an glialen und neuronalen Zellkulturen des Gehirns. Dissertation, University of Tübingen
8. Möller A, Hamprecht B (1989) Creatine transport in cultured cells of rat and mouse brain. Neurochemistry 52:544–550
9. Birken DL, Oldendorf WH (1989) *N*-Acetyl-L-aspartic acid: a literature review of a compound prominent in ^{1}H-NMR spectroscopic studies of brain. Neurosci Biobehav Rev 13:23–31
10. Barany M, Arus C, Yen-Chung Chang (1985) Natural-abundance ^{13}C NMR of brain. Magn Reson Med 2:289–295
11. Bradford HF (1986) Chemical neurobiology. Freeman, New York
12. Jans AWH, Leibfritz D (1989) A C-13 NMR study on fluxes into the Krebs cycle of rabbit renal proximal tubular cells. NMR Biomed 1:171–176

Magnetic Resonance Spectroscopic Evaluation of the Effects of High-Energy Shock Waves Administered to a Human Kidney Cancer Xenograft

A. Heerschap[1], G.A.H.J. Smits[2], H.J. van den Boogert[1], G.O.N. Oosterhof[2], F.M.J. Debruyne[2], J.H.J. Ruijs[1], and J.A. Schalken[2]

Introduction

The administration of high-energy shock waves (HESW) in clinical lithotripsy for kidney stone fragmentation is well known [1], but recent studies have revealed other potential biomedical applications for this type of acoustic energy. It has been known for a few years that HESW can be used to suppress tumor growth in vivo [2]; a temporary suppression of tumor growth for different xenografts after focal administration of HESW has been demonstrated [3]. However, the processes underlying this biological effect are poorly understood. HESW may affect tumor cells and their metabolism both directly and indirectly (for example, at the vascular level).

The purpose of this study was to employ magnetic resonance (MR) as an adjuvant in the elucidation of HESW-induced alterations of tumor tissue processes. ^{31}P MR is established as a useful tool for longitudinal metabolic and physiological studies on high-energy phosphates in in vivo growing tumor xenografts, and we employed this methodology to investigate the effects of HESW on NU-1 human kidney cancer xenografts [4]. In addition, we used ^{1}H MR to monitor the presence of lactate in these tumors and ^{2}H (deuterium) MR to evaluate tumor blood flow (TBF). The latter approach was introduced recently [5] and is based on the observation of the washout of deuterium-labeled water ($^{2}HO^{1}H$) after the administration of a bolus of $^{2}H_2O$ to the tumor.

Materials and Methods

The NU-1 human kidney cancer xenografts were grown subcutaneously in the hind limb of BALB/c nude mice. Tumors (0.2–0.4 cm^3) were exposed to 0, 200, and 800 electromagnetically generated HESW employing a Siemens Lithostar ($P^+ = 37.5$ MPa, $P^- = 5.2$ MPa, 0.75 Hz). MR measurements

[1] Department of Diagnostic Radiology and [2] Department of Urology, University Hospital Nijmegen, P.O. Box 9101, 6500 HB Nijmegen, The Netherlands

were performed on a Bruker spectrometer equipped with a 4.7-T magnet employing home-built $^1H/^{31}P$ and $^1H/^2H$ double-tunable two-turn surface coils. The tumors were lead through a matched hole in a Faraday shield and positioned adjacent to the coil. The mice were anesthetized with a flow of 1.5% enflurane in a O_2/N_2O mixture administered through a nosecone, and their body temperature was maintained by a flow of humidified warm air. ^{31}P spectra were obtained with a composite pulse with a nominal 90^0 flip angle at about 1 mm from the coil centre (TR = 3.5 s). Tumor pH was evaluated from he pH-dependent chemical shift of the inorganic phosphate (Pi) resonance.

^1H MR spectra were obtained with a spin-echo sequence with EXORCYCLE phase depth selection, preceded by radiofrequency pulses and a spoiler field gradient to reduce H_2O and high-flux signals. A hard radiofrequency pulse was used for excitation and a selective 1-1 pulse (maximum at 1 ppm) for refocusing. Spin-echo difference spectroscopy was used for lactate editing [6]. The spin-echo sequence was modified for this purpose with a rotating-phase DANTE [7] to enable selective inversion of the methine spins of lactate. The phases of the DANTE pulses were rotated alternatively clockwise and counterclockwise for on- and off-resonance inversion. Further suppression of unwanted excitation of triglyceride signals was obtained by alternatively placing the DANTE pulses before and after the refocusing pulse [6]. A TE of 152 ms and a TR of 2.5 s were used. A spectrum without H_2O suppression was also recorded with the same spin-echo sequence but avoiding T1 and T2 weighting (TE = 1 ms, TR = 7 s), to obtain the tissue H_2O resonance which was used for absolute quantification of lactate. For ^2H MR spectroscopy 3 × 10 µl ^2H_2O was injected at different sites in the tumor. Within 2 min after the injection serial ^2H MR measurements (64 scans, TR = 0.5 s) were started and continued for about 1 h to follow $^2HO^1H$ washout.

Results and Discussion

^{31}P MR spectra of a NU-1 tumor xenograft obtained at various time intervals before and after the administration of 800 HESW are shown in Fig. 1. After HESW treatment a decrease in nucleoside triphosphate (NTP, mainly ATP) and phosphocreatine PCr resonances concomitant with an increase and acidic shift in the Pi resonance are observed, which are then followed by recovery to the original resonance intensities and positions. For the tumors studied after treatment with 800 HESW (n = 6) the lowest average NTP/Pi ratio was 9% of the starting ratio (range 0%–20%). This value was reached about 6.5 h after treatment. After the administration of 200 HESW (n = 4) the average NTP/Pi ratio dropped to 47% (39%–67%) of its starting value, which was reached after 8 h. The original NTP/Pi ratio was obtained again after about 26 and 52 h for the 200 and 800 HESW groups, respectively. In

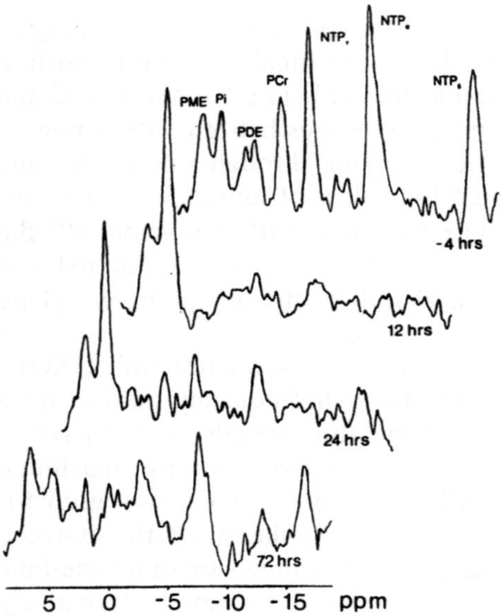

Fig. 1. ^{31}P MR spectra from NU-1 kidney cancer xenografts at various times before and after focal administration of a single dose of 800 HESW. *NTP*, resonances for nucleoside triphosphates; *PCr*, resonance for phosphocreatine; *Pi*, resonance for inorganic phosphate; *PDE*, phosphodiester; *PME*, monoester. (From [4] with permission)

both groups tissue pH decreased from 7.1 to less than 6.8 (at 3–8 h after HESW) and recovered to its original value within the following 2 days. These results stand in contrast to similar ^{31}P MR studies on the effects of HESW on skeletal muscle tissue in which far less dramatic changes were observed [8].

In ^{1}H MR spectroscopy of living tissue the methyl lactate resonance can be used as an important signal to probe anaerobic glycolysis. Unfortunately, in spectra of tumor xenografts this signal coresonates with the dominating signals of triglycerides from subcutaneous adipose tissue. However, the lactate signal can be resolved from these triglyceride signals by special editing radiofrequency pulse sequences. The effectiveness of our editing approach (described above) to monitor lactate in the case of NU-1 kidney cancer xenografts is demonstrated in Fig. 2. The integral of the resolved lactate signal can be compare with that of the tissue water resonance signal, and under certain assumptions the absolute lactate concentration can be estimated from this comparison. Figure 3A shows the pH and lactate evaluation of four combined ^{31}P and ^{1}H MR experiments after the administration of 800 HESW. The rise and decline in lactate tissue concentration follows a similar timing as the decrease and increase in tumor pH.

Both the changes for high-energy phosphates and the correlated variations of lactate levels and tissue pH after HESW are indicative of the

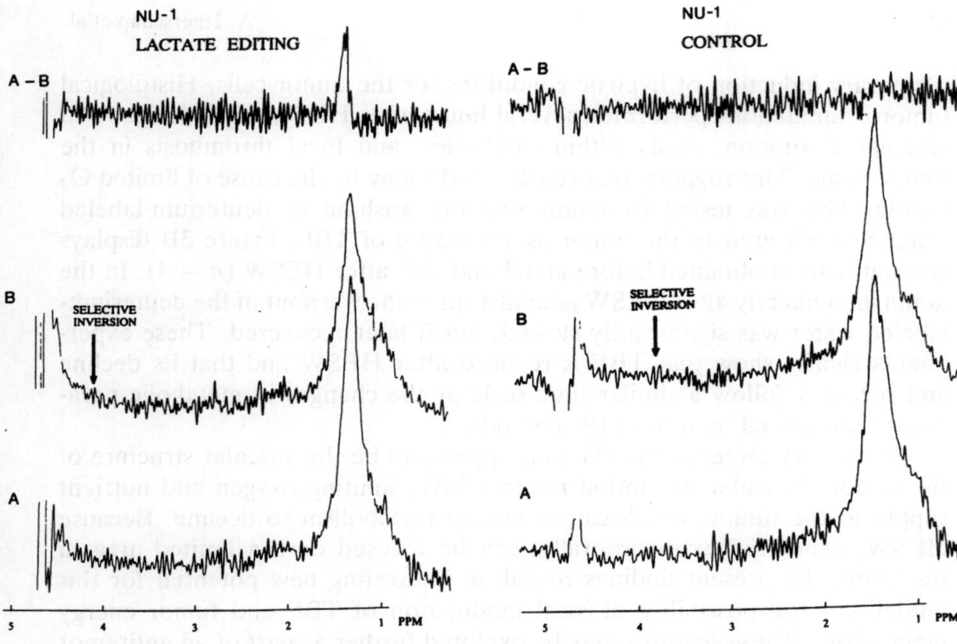

Fig. 2. Lactate editing in in vivo ^1H MR spectroscopy of NU-1 human kidney cancer xenografts to resolve the methylgroup resonance of lactate from resonances of triglycerides. *Left*, lactate editing experiment. Selective inversion by the DANTE pulses placed at 4.1 ppm (*B*) to invert the methine spins of lactate and at −2.1 ppm (*A*) for off-resonance inversion. The lactate signal is obtained at 1.3 ppm in the difference spectrum (*A* − *B*). The large signal (*traces A and B*) is mainly from triglycerides. *Right*, control experiment. No signal is observed at 1.3 ppm in the difference spectrum (*A* − *B*) after shifting the selective inversion to 3.8 ppm, thus confirming that the signal at 1.3 ppm obtained in the previous experiment originates from lactate

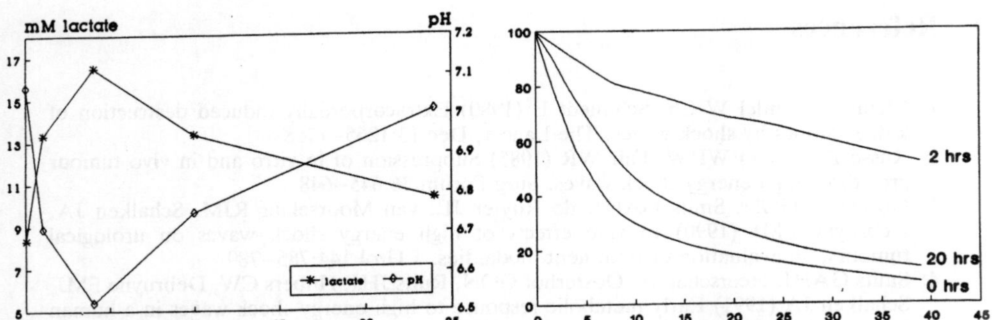

Fig. 3. Longitudinal MR studies of NU-1 human kidney xenografts (n = 4). *Left*, Correlation between variations in tumor lactate concentration and tumor tissue pH after HESW administration. At 0 h are control values. *Right*, Washout curves of $^2HO^1H$ in the tumor after the administration of a bolus of ^2H_2O. The relative integral of the deuterium signal with respect to the starting value is plotted as a function of time. Curves are shown that have been obtained before (0 h) and 2 and 20 h after treatment with 800 HESW

temporary induction of hypoxic conditions for the tumor cells. Histological tumor examinations performed several hours after HESW exposure showed vascular disruption, stasis within capillaries, and focal thrombosis in the tumor tissue. This suggests that reduced TBF may be the cause of limited O_2 supply. This was tested by monitoring the washout of deuterium-labeled water administered to the tumor as a measure of TBF. Figure 3B displays washout curves obtained before and 2 and 20 h after HESW ($n = 4$). In the hours immediately after HESW administration the washout of the deuterium-labeled water was significantly slowed, but it later recovered. These experiments clearly show that TBF is reduced after HESW and that its decline and recovery follow a similar time scale as the changes in metabolic parameters, measured by other MR methods.

A primary target of HESW thus appears to be the vascular structure of the tumor. Vascular disruption reduces TBF, limiting oxygen and nutrient supply to the tumor, which causes energy metabolism to decline. Because HESW, generated extracorporally, can be focused onto a limited area in the body, the present findings reveal an interesting new potential for this modality – the possibility of local modulation of TBF and tumor energy metabolism. These features may be explored further as part of an antitumor treatment strategy.

Acknowledgements. This work was supported by the Dutch Kidney Foundation, the Siemens Corporation (Erlangen, FRG), and the Maurits and Anna de Kock Foundation. The MR spectra were obtained at the Dutch high-frequency MR facility, sponsored by SON (Scheikundig Onderzoek Nederland).

References

1. Chaussy Brendel W Ch, Schmiedt E (1980) Extracorporeally induced destruction of kidney stones by shock waves. The Lancet, Dec 13:1265–1268
2. Russo P, Heston WDW, Fair WR (1985) Suppression of in vitro and in vivo tumour growth by high energy shock waves. Surg Forum 36:645–648
3. Oosterhof GON, Smits GAHJ, de Ruyter JE, van Moorselaar RJM, Schalken JA, Debruyne FMJ (1990) In vivo effects of high energy shock waves on urological tumours, an evaluation of treatment modalities. J Urol 144:785–789
4. Smits GAHJ, Heerschap A, Oosterhof GON, Ruijs JHJ, Hilbers CW, Debruyne FMJ, Schalken JA (1991) Early metabolic response to high energy shock waves in a human tumour kidney xenograft monitored by ^{31}P magnetic resonance spectroscopy. Ultrasound Med Biol 17:791–802
5. Kim SK, Ackerman JJH (1988) Quantitative determination of tumour blood flow and perfusion via deuterium nuclear magnetic resonance spectroscopy in mice. Cancer Res 48:3449–3453
6. Hanstock CC, Bendall MR, Metherington HP, Boisvert DP, Allen PS (1987) Localized in vivo proton spectroscopy using depth-pulse spectral editing. J Magn Reson 71:349–354

7. Blondet P, Albrand JP, von Kienlin M, Decorps M (1987) Use of rotating-phase DANTE pulses for in vivo proton NMR spectral editing with a single irradiation facility. J Magn Reson 71:342–346
8. Smits GHAJ, Jap PHK, Heerschap A, Oosterhof GON, Debruyne FMJ, Schalken JA (1992) Bio-effects of high energy shock waves in mouse skeletal muscle: correlation between ^{31}P magnetic resonance spectroscopic and microscopic alterations. Ultrasound Med Biol

Magnetic Resonance and Kidney Function

^{31}P and ^{1}H Magnetic Resonance of the Kidney: From Experiments in Animals to Applications in Humans

F. Terrier[1,2], F. Lazeyras[1,3], F. Frey[1], E. Scott[1,2], A. Montandon[1], and P. Vock[1]

Introduction

Episodes of renal failure frequently follow kidney transplantation. Most of these are reversible with appropriate medical or surgical treatment. Early and accurate diagnosis of posttransplantation renal failure is the key to better long-term survival of the renal allograft. The need for close monitoring of renal allografts is therefore obvious. Besides surgical complications, acute tubular necrosis, acute rejection, and cyclosporin nephrotoxicity are the most frequent causes of early posttransplantation renal failure. Diagnosing the etiology of early posttransplantation renal failure is not easy, either clinically or by current laboratory and imaging methods. Indeed, it would be of great interest if the clinically important differentiation between acute rejection and cyclosporine nephrotoxicity could rely on precise parameters, for example, the changes in concentration of intrarenal metabolites. This differential diagnosis has direct implications for patient management because it means higher dosage of immunosuppressive agents, on the one hand, or cessation of cyclosporine administration, on the other. Therefore, several authors have expressed the hope that magnetic resonance (MR) spectroscopy can offer measurable and accurate parameters (a) to help establish the etiological diagnosis of posttransplantation renal failure and (b) to evaluate more efficiently the prognosis of a rejection episode so that the decision to continue the treatment or to retransplant can be made on objective grounds.

A few years ago we planned to evaluated the ability of MR spectroscopy to differentiate between the various causes of posttransplantation renal failure. In this chapter we review the experiments that we performed in animals and in humans with this goal in mind. We showed that multinuclear (^{31}P and ^{1}H) MR spectroscopic studies provide a detailed insight into im-

[1] Department of Diagnostic Radiology, MR Center and Polyclinic of Internal Medicine, University of Bern, CH-3010 Bern, Switzerland
[2] Present address: Department of Radiology, University Hospital of Geneva, 24, Rue Micheli-du-Crest, CH-1211 Genève 4, Switzerland
[3] Present address: Department of Radiology, Duke University Medical Center, Durham, NC 27710, USA

portant metabolic events occurring during acute ischemia [20]. We also established, as have others, the feasability of performing MR spectroscopic studies in patients with kidney transplants in a clinical environment on a current 1.5-T whole-body MR imager. However, we are not yet convinced, either on the basis of our own experience or that from published data, that MR spectroscopy will eventually evolve into a clinically useful diagnostic tool for the investigation of posttransplantation renal failure. We therefore indicate the main reasons for our rather critical point of view regarding this application of MR spectroscopy.

^{31}P and ^{1}H MR Spectroscopy in Rats

Because ischemic injuries are involved in acute tubular necrosis, in acute rejection, and probably also in cyclosporine nephrotoxicity, we first studied the effect of acute ischemia on the renal content of high-energy phosphates, pH, and lactate in the rat. During ischemia, depletion of ATP, which is the major energy supply of the cell, eventually leads to irreversible cell damage. In ischemia, the source of ATP is exclusively via anaerobic glycolysis, which converts glucose and glycogen to lactate. Increased lactate production by the kidney results from anearobic glycolysis and is therefore the consequence of impaired oxygenation, mostly due to reduced blood supply. On the other hand, cessation of lactate production during ischemia indicates exhaustion of the substrates of anaerobic glycolysis and is probably linked to the loss of cell vitality.

It is a commonly believed that the drop in pH observed during ischemia is related to an accumulation of lactate. However, it is now recognized that hydrolysis of ATP is the primary mechanism of cellular acidification during anoxia [8]. In a series of experiments, we determined the time course of the change in intracellular pH and the concentration of ATP and lactate. In particular, we studied how closely the drop in pH correlates with the consumption of ATP and the accumulation of lactate.

Because lactate could be a valuable indicator of the severity of ischemia, one would like to be able not only to detect lactate in the kidney but also to obtain quantitative information on its concentration. Therefore, we designed a method for determining the absolute intrarenal lactate concentration [11]. Finally, in a further set of experiments using a chemical shift imaging (CSI) technique, we studied the intrarenal distribution of lactate during renal ischemia [10, 21].

Materials and Methods

Surgical Preparation

Depending on the type of experiments, we used different methods to produce renal ischemia. Adult female Wistar rats were anesthetized with sodium thiopental (20–30 mg/rat intraperitoneally). The left kidney of each rat was exposed through a midline incision. The perirenal fat was removed. The renal vessels were exposed, and the renal artery was dissected free from the renal vein. For experiments designed to study dynamic changes of intrarenal metabolites, we adopted the following setting: a sling was loosely looped around the renal artery and was exteriorized through a polyethylene tubing fixed by two sutures to the paraspinal muscle. In this way the kidney could be made ischemic by pulling on the sling without moving the rats out of the magnet. For lactate quantitation and mapping experiments, ischemia was induced at the time of the surgical preparation. The renal pedicle (including the renal artery and vein) and the ureter (along with the periureteral vessels) were ligated to achieve complete ischemia.

A radiofrequency (RF) coil was placed around the left kidney. The peritoneal cavity was then closed. Abdominal muscles and skin were tightly sutured around the coil leads. The rat was positioned into the magnet so that the left kidney was centered in the middle of the static magnetic field.

Coil Design and MR Spectroscopic Measurements

The RF coil consisted of a Helmholtz pair of insulated copper wires. Its diameter was 1.2 cm. It was inserted axially around the kidney near the hilus. Layers of 1-mm-thick Gortex were placed between the coil and surrounding tissues to reduce signals from outside the coil. The circuitry of the coil was designed so that the coil could be tuned to either ^{1}H or ^{31}P. The MR experiments were performed on a horizontal 22-cm clear-bore in vivo MR system (CSI, General Electrics, Fremont, CA), equipped with a 2.0-T superconducting magnet (Oxford Instruments, Oxford, UK) and a second RF channel (decoupler). Maximum gradient strength was 3 G/cm. The proton signal arising from water was used to shim the magnetic field to a water line width of 25–35 Hz.

For dynamic ^{31}P MR spectroscopic experiments we used the following protocol. After control spectra had been obtained, renal ischemia was induced by pulling on the sling. One-pulse ^{31}P spectra were collected every 2.5 min for 30 min, then every 5 min for the next 30 min. The ^{31}P MR spectra consisted of 128 acquisitions, using a spectral width of ±2000 Hz, a block size of 2 K complex points, and a TR of 1000 ms. In the control spectra the beta peak of ATP was assigned a chemical shift of −16.0 parts per million (ppm). Intracellular pH was determined from the chemical shift of the inorganic phosphate (Pi) peak, according to the equation: $pH = 6.75 + \log(d - 3.27/5.78 - d)$, where d is the chemical shift of Pi.

For lactate spectroscopy, we utilized a double resonance editing scheme [16]. This works as follows: the basic sequence is a water suppression Hahn spin echo, 1331-t-2662-t-ACQ, with the maximum sensitive region set on the CH_3 resonance of lactate (1.3 ppm). This pulse sequence does not modulate the phase of the CH_3 resonance because the coupled CH resonance resides in the null excitation region (4.1 ppm). For the J-modulated echo, a gaussian frequency-selective 180 pulse, set on the CH resonance, was irradiated on the decoupler channel simultaneously with the 2662 pulse. All the echoes had a TE of 136 ms (1/J). The interpulse delay of the 1331 sequence was 1670 μs, the spectral width ±1000 Hz, the number of complex points, 1 K, the width of the selective 180° pulse, 40 ms, corresponding to a 85-Hz excitation bandwidth.

It was important to confirm the identity of the lactate peak because alteration in the mobility of lipids during ischemia has been considered responsible for the appearance of a peak at the chemical shift of lactate. Thus, we extended the modulation experiment to TE values of 272 and 408 ms. Because the peak at 1.3 ppm gave maximum signal by using a TE of 136 ms (TE = 1/J), null signal by using a TE of 272 ms (TE = 2/J), and a T2-weighted signal by using a TE of 408 ms (TE = 3/J), it was indeed due to lactate and not to improperly suppressed lipids. We obtained sequentially 1-min ^{1}H edited MR spectra during ischemia with TR = 1000 ms, and the number of acquisitions (NA)-64. They were collected every 2.5 min for 30 min and then every 5 min for the next 30 min. The carrier frequency was set on the water peak and it was checked before each new acquisition.

We adressed the problem of quantitation by comparing the intensity of the CH_3 lactate peak to the intensity of the water peak, considering the water content of the kidney stable during ischemia. The lactate spectra were obtained using the method described above after 1 h of renal ischemia to ensure a steady state in lactate production. The acquisition parameters were the following: TR, 3 s; NA, 128; spectral width ± 1000 Hz; number of complex data points, 2 K; gaussian pulse duration, 40 ms (corresponding to a 85 Hz-excitation bandwidth), gaussian pulse peak power, 1.2 mW. In addition, the refocusing pulse was phase cycled, which further reduced the water contamination of the spectra. The water signal was acquired with a conventional spin echo using the following parameters: TR, 10 s; TE, 0.5 ms; NA, 8. Under these conditions, we did not expect any influence of T1 and T2 effects on the intensity of the water signal.

To calculate the absolute lactate concentration, we used the following equation: [lactate] = 2/3 × [water] × l(lactate)/l(water) × G × NA/na × f(T1, T2) × p × w; where [lactate] and (water) are the concentrations of lactate and water (the latter being 55.5 mol/l); the factor 2/3 reflects the difference in the number of protons per molecule of water and lactate respectively; l(lactate) and l(water) represent the integrals of the lactate and the water resonance, respectively; G is the receiver gain ratio of the lactate and water acquisitions obtained after proper calibration of the spectrometer,

NA (= 8) is the number of acquisitions of the water spectrum; na (= 128) is the number of acquisitions of the lactate spectrum; *f*(T1, T2) is the correction factor for T1 and T2 effects on the intensity of the lactate peak; *p* is the efficiency of the lactate editing scheme (which was measured to be 60%, see below); *w* is the water content of ischemic kidneys, which we previously determined to be 80%.

For T1 measurements of lactate, an initial inversion pulse was added to the basis sequence. A total of 12 inversion intervals (TI) was used for the T1 estimation by a three-parameter fit. For T2 measurements, we used the signal obtained at TE = 136 ms and the one obtained at TE = 408 ms. The T1 and T2 of renal lactate measured in vivo in eight rats were the following: 1350 ± 140 ms and 165 ± 8 ms, respectively. The efficiency reflects the doublet inversion quality under in vivo conditions. It was defined as the ratio of the edited lactate peak area to 2 × the area of the spectrum obtained from the nonmodulated echo (the factor 2 comes from the fact that the signal intensity of the edited spectrum is double that of the nonedited spectrum). This measurement was possible only at TE = 408 ms, when the fat peak was considerably reduced.

For mapping of intrarenal lactate, we used a four- and a three-dimensional CSI sequence [4]. Spatial localization was achieved for the four-dimensional CSI by three orthogonal phase-encoding gradients applied after the gaussian pulse. In three-dimensional CSI, phase-encoding gradients were employed only for localization in the axial plane of the kidney, whereas localization along the longitudinal axis of the kidney was determined by the sensitivity profile of the Helmholtz coil. Acquisition parameters were the following: TR, 1000 ms; TE 136 ms; field of view (FOV) in the axial plane of the kidney 30 mm, with 16 phase-encoding steps; NA, 16 per phase encoding step with a 8-step phase cycling. For four-dimensional CSI experiments: FOV along the longitudinal axis of the kidney, 30 mm with 8-phase encoding steps; matrix size, 16 × 16 ×8; voxel size, 1.87 × 1.87 × 3.50 mm^3 (12.3 µl). Total acquisition time was 9 h 11 min. For three-dimensional CSI experiments: matrix size, 16 × 16; voxel size, 1.87 × 1.87 mm^2 in the axial plane and approximately 14 mm in the longitudinal axis of the kidney as determined by the sensitive volume of the Helmholtz coil. Total acquisition time was 1 h 8 min.

Immediately after the MR measurements had been performed, the left kidney of each rat was removed for enzymatic determination of lactate.

Results and Discussion

On the control ^{31}P MR spectra (Fig. 1), the peaks observed with their chemical shift were congruous with results of previous studies. A small but distinct phosphocreatine (PCr) peak was seen in some of the rats, due either to a slight contamination from surrounding abdominal wall muscles or from muscle tissue in the wall of the intrarenal vessels. Immediately after occlusion

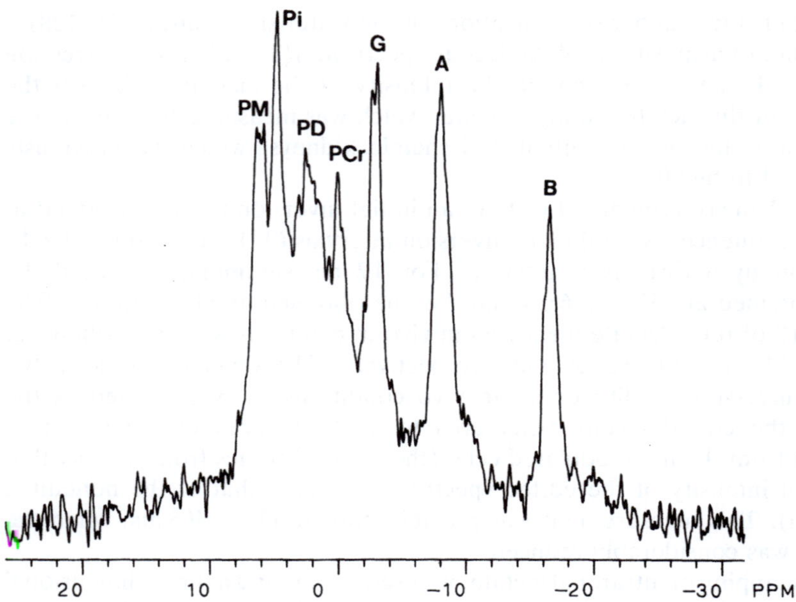

Fig. 1. ^{31}P spectrum of a rat kidney acquired using an implanted coil. The acquisition parameters were: TR, 1000 ms; NA, 600. *PM*, phosphomonoesters; *Pi*, inorganic phosphate; *PD*, phosphodiesters; *PCr*, phosphocreatine; *G, A, B*, gamma, alpha and beta peaks of ATP. (From [20] with permission)

of the renal artery the Pi peak increased progressively whereas the three peaks of ATP decreased (Fig. 2). The drop in beta ATP was characterized by a time constant of 1.9 ± 1.2 min, assuming an exponential decay. The mean intracellular pH fell rapidly below 7.0 within 1 min of ischemia to reach a plateau around 6.7 after about 5–10 min of ischemia. Its time constant was 2.3 ± 1.3 min.

In the control-edited ^1H spectra, there was either no detectable lactate peak or a very small one corresponding to a maximal concentration of about 2 μmol/g. Following occlusion of the renal artery, a lactate peak was observed within 1–6 min of ischemia in all rats, steadily increasing in intensity for 1 h (Fig. 3). The mean time constant was 15.9 ± 7.5 min assuming an exponential increase, which means that 99% of the maximum concentration of lactate would have been reached after 80 ± 38 min ($5 \times$ time constant). The time constant of lactate accumulation was significantly longer than that of the pH drop ($p < 0.005$). There was no significant difference between the time constant of the decline in ATP concentration and that of the pH drop.

Our MR results on the absolute intrarenal lactate concentration were within 20% of the results of chemical analysis. In eight rats the lactate concentration measured by MR was 12 ± 1 μmol/g wet tissue, and it was 13 ± 1 μmol/g wet tissue measured chemically. Using ^1H CSI, we observed

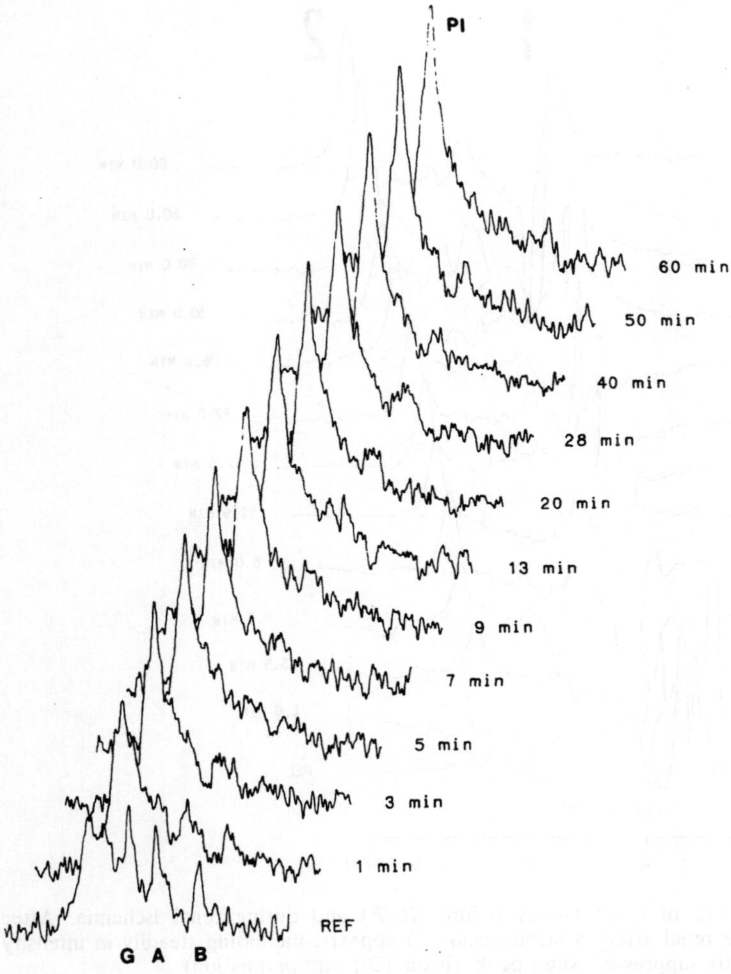

Fig. 2. ^{31}P spectra of a rat kidney before (*REF*) and during renal ischemia. There is a rapid increase in Pi and a decrease in the three peaks of ATP. (From [20] with permission)

that the lactate accumulated mostly in the renal medulla during ischemia (Fig. 4). This was verified by chemical analysis which showed that lactate concentration was 17 ± 4 μmol/g wet tissue in the medulla whereas it was 9 ± 4 μmol/g wet tissue in the cortex ($p < 0.001$). In summary, our experiments showed that in the kidney the decrease in intracellular pH does not correlate with the production of lactate but with the hydrolysis of ATP, according to the diagram in Fig. 5. They also demonstrated that it is possible to obtain reliable data on the absolute lactate concentration with MR ^{1}H spectroscopy. However, this task is difficult, and a careful analysis of the

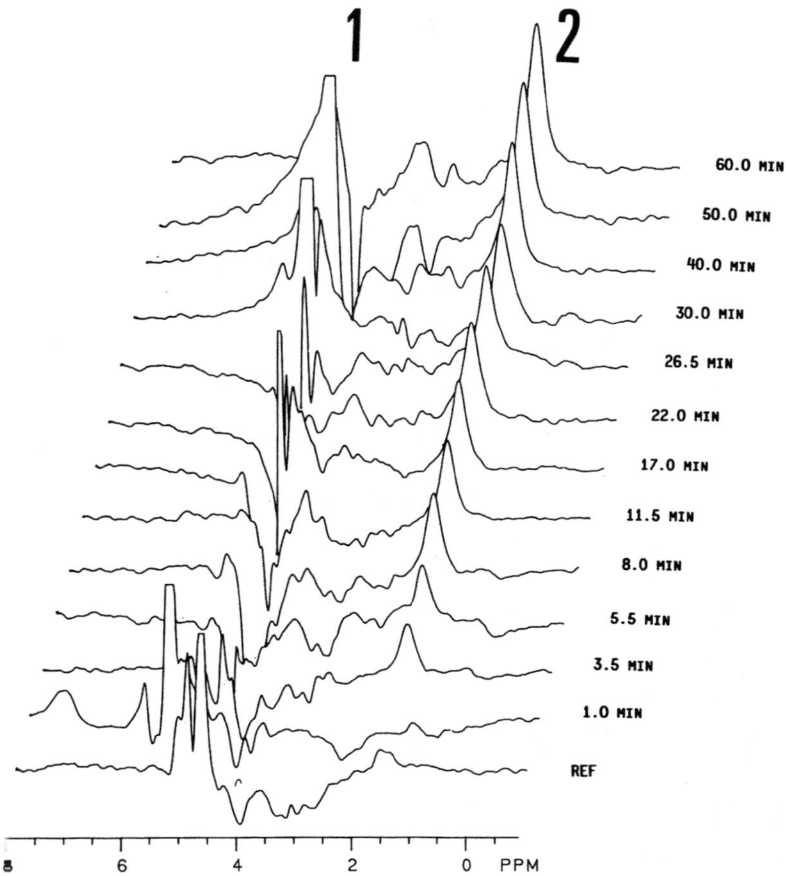

Fig. 3. ^{1}H spectra of a rat kidney before (*REF*) and during renal ischemia. After occlusion of the renal artery a lactate peak (*2*) appears, increasing steadily in intensity over 1 h. *1*, Partly suppressed water peak. (From [20] with permission)

method used is necessary. The good correlation between MR and CSI results indicates further that under our experimental conditions all the renal lactate was visible, and that therefore no lactate pool undetectable by MR spectroscopy. Finally, the lactate images that we acquired depict the uneven intrarenal distribution of lactate during ischemia, this metabolite accumulating mostly in the medulla.

Encouraged by theses results and those obtained by Shapiro et al. [18] showing in rats that ^{31}P MR spectroscopy can differentiate among the various causes of posttransplantation renal failure, namely ischemic injury, acute rejection, and cyclosporin toxicity, our next goal was to implement the necessary hard- and software on our whole-body 1.5-T magnetic resonance

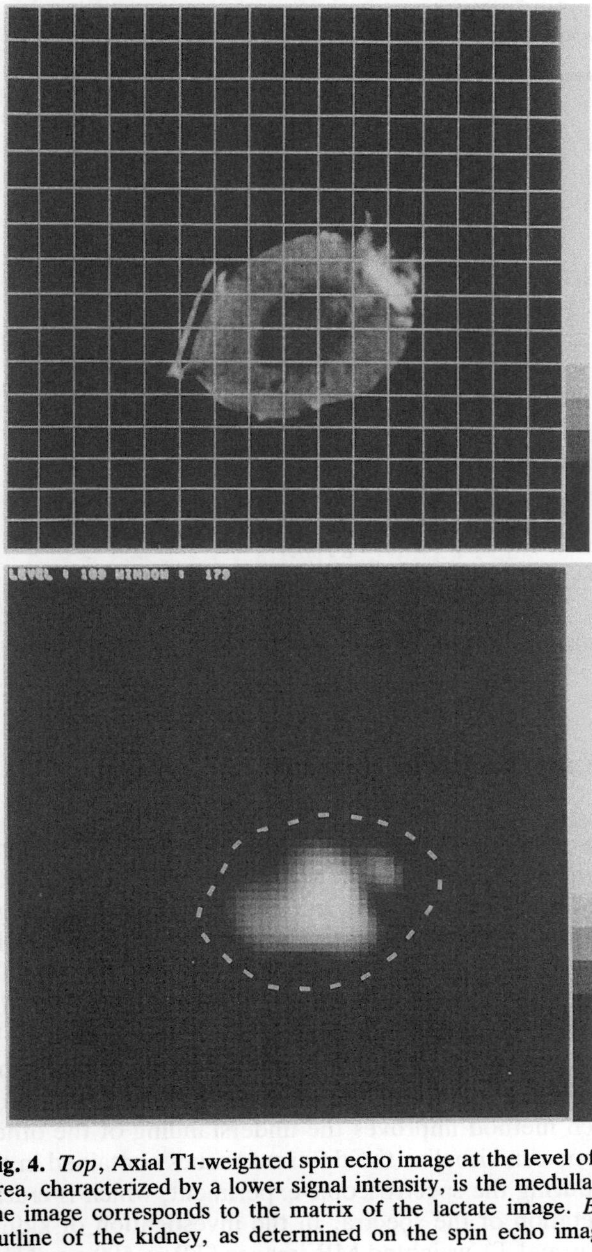

a

b

Fig. 4. *Top*, Axial T1-weighted spin echo image at the level of the renal hilus. The central area, characterized by a lower signal intensity, is the medulla. The grid superimposed on the image corresponds to the matrix of the lactate image. *Bottom*, Lactate image. The outline of the kidney, as determined on the spin echo image, is drawn on the lactate image

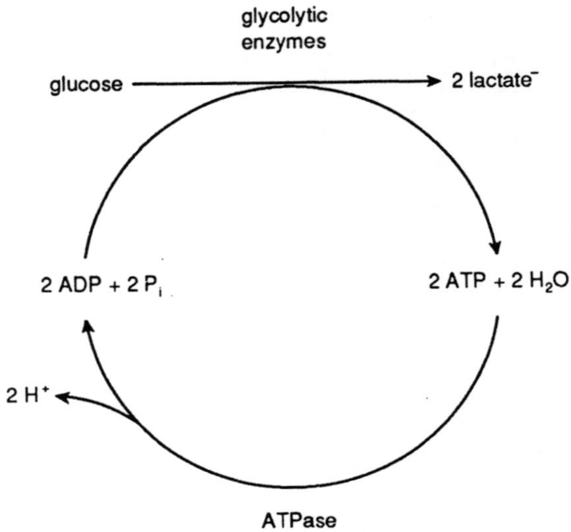

Fig. 5. Diagram showing the relationship between glycolysis and ATP hydrolysis

system to perform localized ^{31}P and ^{1}H MR spectroscopy of human kidney transplants.

MR Spectroscopic Experiments in Humans

The MR measurements were performed on a whole-body 1.5-T magnet (Signa, General Electric Medical Systems, Milwaukee, WI, USA). This system is located in a hospital environment and used for clinical imaging 4 days a week and for clinical spectroscopy 2 days a week. It is equipped with the spectroscopic research package of GE and employs 1 G/cm shielded gradient coils with eddy-current compensation. Our strategy in performing spectroscopy in humans is the "integrated approach" as recommended by Lenkinski [12], combining imaging and spectroscopy in a single examination. By acquiring MR images and spectra in the same examination, the information gained from each method improves the understanding of the other. Thanks to the images, one can visualize the different tissues contained in the volume being sampled during the spectroscopic experiment, which is crucial for an accurate interpretation of the spectra. In the investigation of kidney transplants, axial and coronal T1-weighted MR images (TR = 500 ms, TE = 10 ms) allow the localization of the transplant within the patient's body and the selection of the best sampling volume for the spectroscopic part of the examination, minimizing contamination from adjacent musculature.

Furthermore, modifications of the corticomedullary contrast on these images is a useful criterion for assessing the presence of acute rejection [9].

^{31}P MR Experiments in Humans

Subjects and Methods

Thirteen patients with kidney transplants were included in the study. In 12 patients, the transplant functioned well whereas in one patient it presented an episode of acute rejection, verified by clinical and laboratory parameters and which was successfully treated by appropriate steroid therapy. Each patient was placed supine. The position of the kidney transplant in the body was determined by coronal and axial spin echo images. A 12.5-cm circular surface coil double-tuned for phosphorus at 25.85 MHz and protons at 63.85 MHz was placed on the abdominal wall over the transplant kidney. The static magnetic field was shimmed without localization scheme for the proton signal arising from water to a linewidth of 20–35 Hz. To obtain localized ^{31}P spectra we used a depth-resolved surface coil spectroscopy (DRESS) sequence [3] in 11 patients and a one-dimensional CSI sequence [4] in three. For the DRESS experiments, acquisition parameters were the following: TR, 1500 ms; TE, 20 ms; slice thickness, 30 mm; data size, 256 complex points; spectral width, 2000 Hz; NA, 512; total acquisition time, 12.8 min. For the one-diminsional CSI experiments: FOV, 960 mm; number of phase encoding steps, 32; NA, 16 per phase encoding step, yielding the same acquisition time of 12.8 min. The whole experiment, including imaging and spectroscopy did not last longer than 90 min.

Results and Discussion

Figure 6 shows a typical ^{31}P MRS spectrum of a normal kidney transplant. The three peaks from ATP as well as those from the phosphomonoesters (PME), the Pi, and the phosphodiesters (PDE) are easily detected. In all patients the PCr peak was high, indicating important contamination by surrounding musculature. Using the one-dimensional CSI technique, contamination was smaller than with the DRESS technique. Furthermore, the baseline distortion ("rolling") produced in the resulting spectra by the delay between the center of the RF pulse and the start of the data acquisition was less disturbing with the CSI than with the DRESS technique. In addition, one-dimensional CSI is a multivoxel technique and allows for retrospective selection of the voxels. In this way, one can minimize a posteriori contamination by surrounding tissues.

In the patient presenting acute rejection, the ^{31}P MR spectra showed no changes in comparison with spectra obtained in normally functioning transplant. We attributed this disappointing result to an inefficient localization

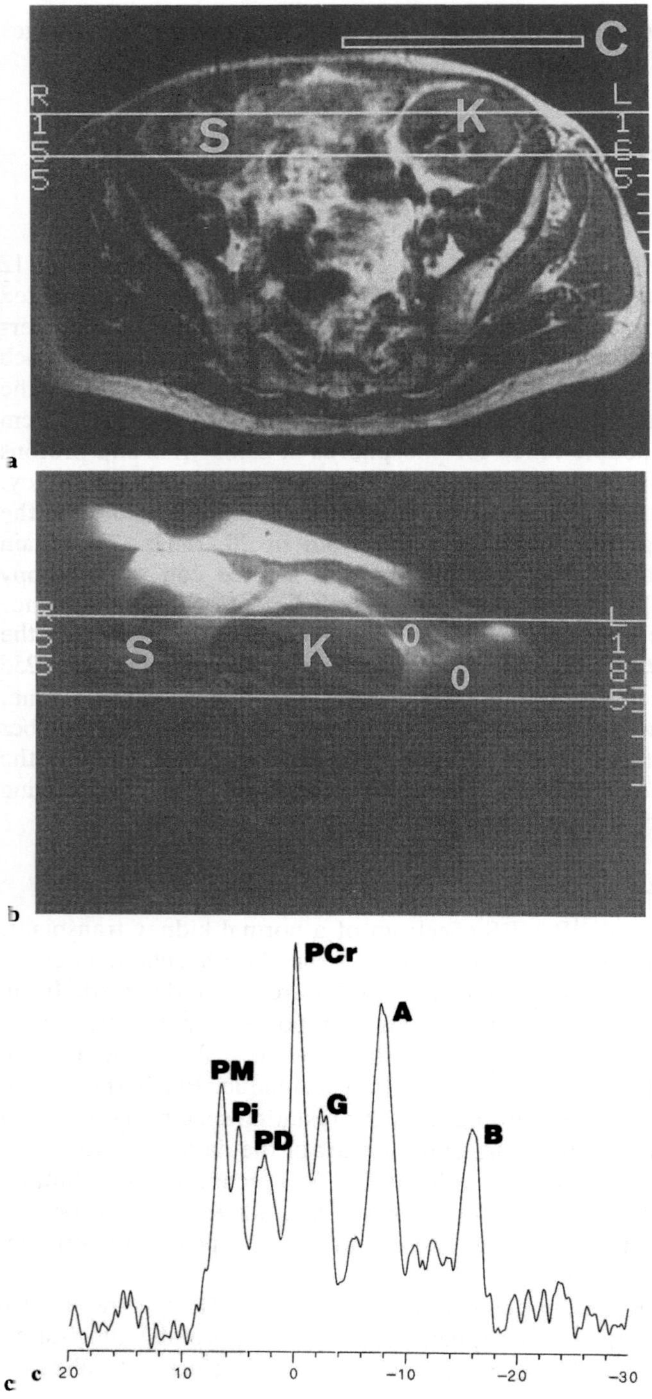
C
R 155
S
K
L 165
a
S
K
0
0
L 185
b
PCr
A
PM
Pi
PD
G
B
20
10
0
−10
−20
−30
c

scheme. A complete volume localization scheme, like a three-dimensional CSI or a two-dimensional CSI in combination with DRESS, should be used to improve the localization and so minimize the contamination by nonrenal tissue. In addition, the data suffered from low sensitivity due to the position of the kidney transplant far from the surface coil. An array coil wrapped around the abdomen and providing a more homogeneous sensitivity profile should increase the signal-to-noise ratio.

^{1}H Experiments in Humans

Subjects and Methods

We studied six patients, all with a normally functioning kidney transplant (Fig. 7a). A 12.5-cm circular surface coil was placed on the abdominal wall just above the kidney transplant. A volume of interest of $2 \times 2 \times 2$ cm was chosen inside the kidney transplant from the spin echo images. Given the size of the voxel, the tissue under investigation included both renal cortex and renal medulla. A stimulated echo acquisition mode (STEAM) localization sequence [5] was used for shimming and acquisition of the ^{1}H spectra (Fig. 7b). The spectra were acquired using the body coil to transmit and the surface coil to receive. Water suppression was achieved by chemical-shift selective RF pulses in conjunction with dephasing gradients, preceeding the STEAM sequence. TR was 2000 ms. One or more values of TE were used in each patient (6, 7, 20, 30, 136, or 270 ms). NA was usually 256 (Fig. 7c).

Results and Discussion

The ^{1}H spectra obtained with TE of 20 ms gave the best signal-to-noise ratio. Shorter TE values resulted in significant degradation of the spectrum from the effects of eddy currents induced by the switching of the gradients. On the other hand, with a TE of 270 ms, the spectra were still of good quality. In all kidney transplants we noted one or more peaks at the chemical shift of metabolites, which according to previous studies in animals and in humans have been identified as so-called osmolites [1, 17, 22]. These metabolites are sequestrated in the cells of the renal medulla and, thanks to

←

Fig. 6. a T1-weighted axial image of a human kidney transplant (*K*) obtained using a body coil. The position of the surface coil (*C*) used to acquire the image and the ^{31}P spectrum in **a** and **b**, respectively, as well as that of the excitation slab (*S*) localized with the DRESS technique are indicated. **b** Image of the same kidney transplant as in **a** obtained using a surface coil. The sampling volume for the localized ^{31}P spectrum shown below results from the intersection of the excitation slab (*S*) and the sensitive volume of the surface coil. The amount of muscle tissue (*O*) included in the sampling volume can be estimated. **c** Localized ^{31}P spectrum of the human kidney transplant imaged in **a** and **b**, acquired using the DRESS technique. The acquisition parameters were: TR, 1000 ms, NA, 512; thickness of the excitation slab, 30 mm. The PCr peak indicates significant contamination of the transplant spectrum by muscle tissue

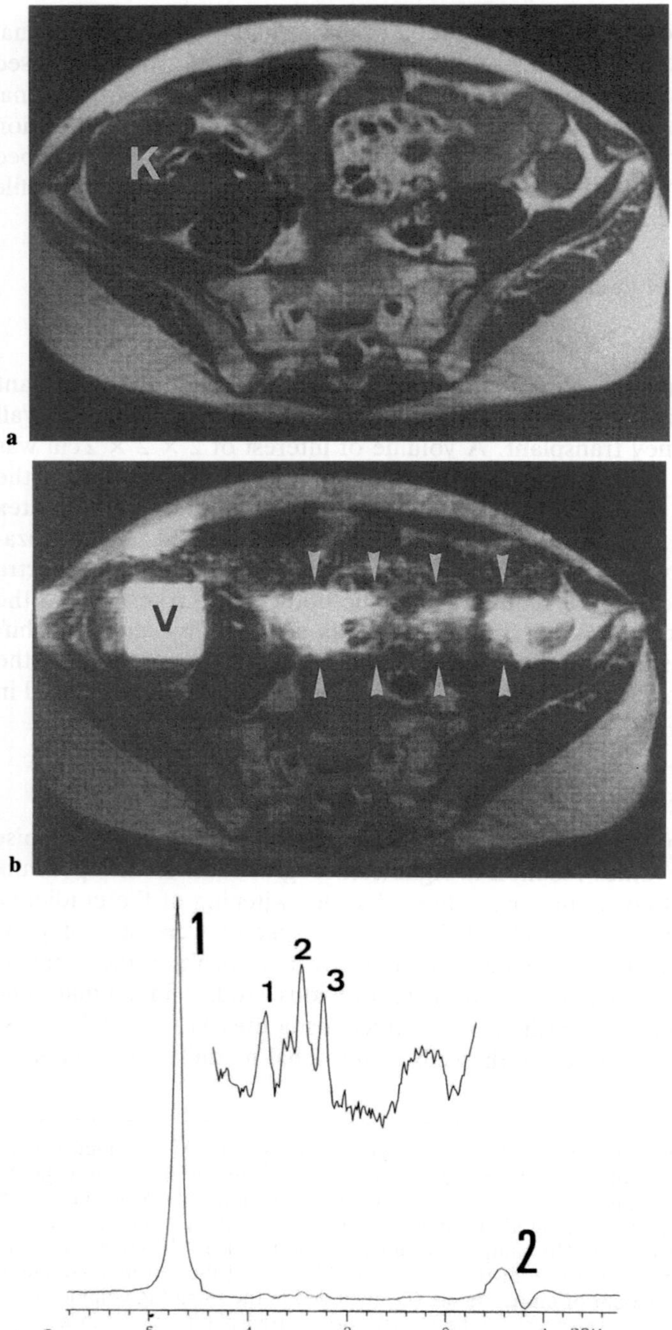
K
V
1
1
2
3
2
5
4
3
2
1
PPM
a
b
c
c

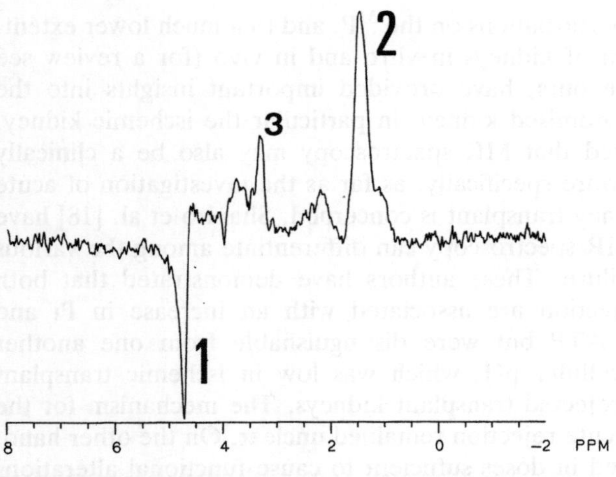

Fig. 8. Localized ^{1}H spectrum of another normally functioning human transplant kidney. Note the important differences in the relative concentration of osmolites in this spectrum in comparison to that in Fig. 7, despite the same acquisitions parameters

their osmotic effect, allow these cells to cope with the highly hypertonic osmolality of the extracellular fluid. They include the trimethylamines betaine and glycerophosphorylcholine and the polyols sorbitol and inositol.

The number of peaks corresponding to the various osmolites as well as their relative concentration varied considerably from one patient to the other (Fig. 8). The only peak which was consistently demonstrated was located at 3.27–3.30 ppm and corresponded to glycerophosphorylamine. Other peaks detected in some patients were assigned to sorbitol and possibly inositol.

Conclusions

As pointed out by Radda et al. [15], "the major long-term benefit from MRS must come from its contribution to the basic biochemical understanding of disease." An extensive amount of work has been carried out to evaluate

←

Fig. 7. **a** T1-weighted spin echo image of a normally functioning human kidney transplant (*K*). **b** The sampling volume (*V*) for localized ^{1}H spectroscopy obtained using the STEAM technique is superimposed on the spin echo image. *Arrow-heads*, artifacts due to breathing. **c** Localized ^{1}H spectrum of the human kidney transplant shown in **a** and **b**. Acquisition parameters were: TR, 2000 ms; TE, 20 ms; NA, 256; sampling volume, 8 cm^3. *1*, Water; *2*, lipids. *Inset*, peaks are those situated between 3 and 4 ppm. According to published data, they probably correspond to: *1*, sorbitol; *2*, inositol; *3*, glycerophosphorylcholine

the effect of metabolic perturbations on the ^{31}P, and to a much lower extent, the ^{23}Na and ^{1}H spectra of kidneys in vitro and in vivo (for a review see [2]). These studies, like ours, have provided important insights into the metabolism of the compromised kidney, in particular the ischemic kidney. They also have suggested that MR spectroscopy may also be a clinically useful diagnostic tool. More specifically, as far as the investigation of acute renal dysfunction of kidney transplant is concerned, Shapiro et al. [18] have shown in rats that ^{31}P MR spectroscopy can differentiate among the various causes of transplant failure. These authors have demonstrated that both ischemia and acute rejection are associated with an increase in Pi and PME and decrease in ATP but were distinguishable from one another by differences in intracellular pH, which was low in ischemic transplant kidneys but normal in rejected transplant kidneys. The mechanism for the absence of pH drop in acute rejection remained unclear. On the other hand, cyclosporine administered in doses sufficient to cause functional alterations in glomerular filtration and renal flow did not cause changes in the relative concentrations of phosphorus metabolites in this animal model of a kidney transplant. However, several other groups have undertaken similar studies in humans and have obtained good-quality spectra of kidney transplants [14].

Some studies have included transplants with acute rejection and have shown changes similar to those observed in experimental investigations [6]. However, in at least one case of confirmed acute rejection, we were not able to demonstrate any change in the ^{31}P spectrum. This finding suggests to us that ^{31}P MR spectroscopy may be an insensitive method, allowing one only at an advanced stage to detect changes in the concentration of renal metabolites during ischemia or acute rejection. An experimental study in rats by Shapiro et al. [19] seems to confirm our preliminary impression. These authors observed modifications in the ^{31}P MR spectrum of rejected kidney transplants at day 7 after transplantation, at a time when a substantial amount of parenchymal tissue injury had already occurred. However, at day 4, no change could be seen by MR spectroscopy, although histological analysis showed that interstitial inflammation was already present. From this study it seems unlikely that ^{31}P MR spectroscopy will permit early detection of acute rejection. In this context, lack of sensitivity appears a major drawback because early treatment of acute rejection is the key to improved long-term survival of the kidney transplant. However, in a recently published paper Grist et al. [7] have concluded, on the basis of a prospective study in 36 patients, that ^{31}P MR spectroscopy may be useful for evaluating renal metabolism during episodes of transplant kidney dysfunction. Obviously, there is a serious need for more clinical studies in humans aimed at defining sensitivity and specificity of ^{31}P MR spectroscopy in acute kidney transplant rejection.

^{1}H MR spectroscopy has contributed greatly to the discovery of a whole class of compounds, the so-called osmolites, which play a major role in protecting the cells of the renal medulla from the extremely high osmolality of the extracellular fluid. However, more experimental work in animals must be performed to establish whether changes in the concentration of these metabolites have any diagnostic significance in diseases. In particular, it will be very interesting to determine what happens to these metabolites in cases of a dysfunction of the urine concentrating mechanism of the kidney. Concerning the feasibility of detecting lactate in human kidney transplants as an indicator of ischemia, we are rather sceptic. Indeed, motion and the presence of large amount of peri- and intrarenal fat are most probably prohibitive for editing techniques.

Thus, although steady and incremental progress is being made, and the quality of localized spectra obtained on clinical whole-body scanners has shown major improvements in recent years [13], clinical MR spectroscopy is advancing slowly and has yet to prove its diagnostic utility. This is true in particular for the investigation of posttransplantation renal failure.

References

1. Bagnasco S, Balaban R, Bales HM, Yang Y-M, Burg M (1986) Predominant osmotically active organic solutes in rat and rabbit renal medullas. J Biol Chem 261: 5872–5877
2. Balaban RS (1989) MRS of the kidney. Invest Radiol 24:988–992
3. Bottomley PA, Foster TB, Darrow RD (1984) Depth-resolved surface coil spectroscopy (DRESS) for in vivo ^{1}H, ^{31}P, and ^{13}C NMR. J Magn Reson 54:338–342
4. Brown TR, Kincaid BM, Ugurbil K (1982) NMR chemical shift imaging in three dimensions. Proc Natl Acad Sci USA 79:3523–3526
5. Frahm J, Merboldt K-D, Hänicke W (1987) Localized proton spectroscopy using stimulated echoes. J Magn Reson 72:502–508
6. Grist TM, Charles HC, Herfkens RJ (1988) P-31 MR spectroscopy of renal transplant recipients with chemical shift imaging. Radiology 169(P):168(abstr)
7. Grist TM, Charles HC, Sostman HD (1991) Renal transplant rejection: diagnosis with ^{31}P MR spectroscopy. Am J Roentgenol 156:105–112
8. Hochachka PW, Mommsen TP (1983) Proton and anaerobiosis. Science 219:1391–1397
9. Hricak H, Terrier F, Demas BE (1986) Renal allografts: evaluation by MR imaging. Radiology 159:435–441
10. Layzeyras F, Aue WP (1990) Lactate distribution in ischemic rat kidney by 4D spectroscopic imaging. NMR Biomed 2:230–233
11. Lazeyras F, Aue WP, Terrier F, Frey F (1993) In vivo determination of absolute lactate concentration (in preparation)
12. Lenkinski RE (1989) Clinical magnetic resonance spectroscopy: a critical evaluation. Invest Radiol 24:1034–1038
13. Luyten PR, Brutink G, Sloff FM et al. (1989) Broadband proton decoupling in human ^{31}P NMR spectroscopy. NMR Biomed 1:177–183
14. Matson GB, Twieg DB, Karczmar GS (1988) Application of image-guided surface coil P-31 MR spectroscopy to human liver, heart, and kidney. Radiology 169:541–547
15. Radda GK, Rajagopalan B, Taylor DJ (1989) Biochemistry in vivo: an appraisal of clinical magnetic resonance spectroscopy. Magn Res Q 5:122–151
16. Rothman DL, Arias-Mendoza F, Shulman GI, Shulman RG (1984) A pulse sequence for simplifying H-1 NMR spectra of biological tissues. J Magn Res 60:430–436

17. Shah NJ, Carpenter TA, Wilkinson ID (1991) Localized in vivo proton spectroscopy of the human kidney. Magn Reson Med 20:292–298
18. Shapiro JI, Haug CE, Weill R III, Chan L (1987) ^{31}P nuclear magnetic resonance study of acute renal dysfunction in rat kidney transplants. Magn Reson Med 5:346–352
19. Shapiro JI, Haug CE, Shanley PF, Weil R III, Chan L (1988) ^{31}P nuclear magnetic resonance study of renal allograft rejection in the rat. Transplantation 45:17–21
20. Terrier F, Lazeyras F, Posse S, Aue WP, Zimmermann A, Frey BM, Frey FJ (1989) Study of acute renal ischemia in the rat using magnetic resonance imaging and spectroscopy. Magn Reson Med 12:114–136
21. Terrier F, Lazeyras F, Frey BM, Frey FJ (1992) Lactate mapping in ischemic rat kidneys using ^{1}H spectroscopic imaging. Invest Radiol 27:282–286
22. Wolff SD, Balaban RS (1987) Proton NMR spectroscopy and imaging of the rabbit kidney in vivo, at 4.7 tesla. J Magn Reson 75:190–192

In Vivo ^{19}F MR Spectroscopy in the Study of Organic Acid Excretion by the Rat Kidney*

U. Ammer[1] and H.P. Juretschke[2]

Introduction

In traditional clearance studies the urinary excretion of a test substance is measured in relation to its plasma concentration (clearance = UV/P; where V is the urine volume per minute, U the urinary concentration of the test substance, and P the corresponding plasma concentration). If the clearance of inulin, which indicates the glomerular filtration rate, is determined simultaneously, it is possible, while accounting for the binding of the test substance to plasma proteins, to evaluate the filtered amount of the test substance. From the filtered and excreted amounts of a test substance it is possible to calculate how much was secreted or reabsorbed. However, traditional clearance studies give no information on the actual intracellular content of a test substance. Using in vivo ^{19}F magnetic resonance (MR) spectroscopy the intracellular concentration of fluorinated organic acids can be monitored continuously. The ^{19}F nucleus has a nuclear spin of 1/2 and, together with its natural abundance of 100%, offers a high detection sensitivity (83% of that of ^{1}H). As biological systems do not normally contain fluorine, no problems with endogenous ^{19}F signals occur.

During secretion or reabsorption in the proximal tubule the substrates must cross the luminal and contraluminal membranes of the tubular cells, and a certain intracellular concentration of the test substance builds up. If substances that interfere with the respective transport system are applied simultaneously, *cis*inhibition or *trans*stimulation of transport may occur. Depending on the direction of transport (secretion or reabsorption) and depending on the mode of interference (*cis*inhibition or *trans*stimulation) at either cell side the intracellular concentration of the test substance rises or falls. When, for example, the secretory transport of a test substance is inhibited, its urinary excretion falls. If inhibition occurs at the contraluminal

---

* Preliminary accounts of this work were presented at the 7th and 8th Annual Meeting of the Society of Magnetic Resonance in Medicine, San Francisco 1988 and Amsterdam 1989.

[1] Max Planck Institut für Biophysik, Kennedyallee 70, D-60596 Frankfurt, FRG
[2] Bruker Medizintechnik GmbH, Silberstreifen, D-76287 Rheinstetten, FRG

cell side, the intracellular concentration of the test substance is expected to fall. However, if inhibition of secretory transport takes place at the luminal cell side, the intracellular concentration of the test substance is expected to rise. However, if secretory transport is *trans*stimulated, the urinary excretion of the tested substance rises. *Trans*stimulation at the luminal side leads to a decline in intracellular concentration of the tested substance, whereas *trans*stimulation at the contraluminal side increases the intracellular concentration of the tested substance. In the case of reabsorptive transport processes similar discrimination for the events at the two cell sides is possible by monitoring the concomitant changes in urinary excretion and tissue concentration.

Appropriate information can be obtained only from ^{19}F MR spectroscopy of renal tissue if urinary excretion and clearance are measured simultaneously. However, it is very difficult to perform clearance studies under the conditions where ^{19}F MR spectra are recorded. The reasons are as follows: (a) for each blood determination the animal must be transiently removed from the magnet; (b) the amount of blood which can be taken from a small animal such as the rat and the low concentration of ^{19}F-labeled test substance make it impossible to measure blood concentrations with MR spectroscopy. Therefore, a compromise has been made in the present study: a single injection procedure without collection of blood samples was used. In this case, urinary excretion and concentration of the test substance in the renal tissue can be related only to the amount of substance injected or to the amount remaining in the organism at a given time.

Several substances were synthetized to study the renal excretion and tissue content with ^{19}F MR spectroscopy: the fluorinated *para*-aminohippuric acid (PAH) derivative 3-trifluoromethylhippurate (TFMH) for investigation of the PAH transport system, tetrafluorosuccinate (TFS) for the succinate transport system, and trifluoroethylsulfate (TESULF) for the sulfate transport system. None of the three substances are metabolized in rats. The apparent K_i values of these substances against the contraluminal anion transport systems were determined in stop-flow microperfusion studies [3].

Methods

^{19}F MR Spectroscopy

The measurements were carried out on a Bruker Biospec 47/40 spectrometer (Bruker Medizintechnik, Karlsruhe, FRG) at a magnetic field strength of 4.7 T corresponding to a frequency of 188 MHz. To increase sensitivity a saddle coil with 10-mm diameter was used into which the exposed kidney was inserted. Usually, 64–240 scans were collected within 1–2 min (relaxation delay of about 0.4 s, acquisition time 0.1 s). ^{19}F MR spectra with adequate signal-to-noise (S/N) ratio were obtained by applying hard pulses with a flip angle of about 70°. The measured free induction decays were Fourier-transformed after zeroing the first two points (in order to obtain a satisfactory base line) and upon applying an exponential multiplication of 70 Hz.

The concentration of the ^{19}F-labeled test substance in urine samples and calibration standards was determined by ^{19}F MR spectroscopy on Bruker AM 250 or AM 300 spectrometer (Bruker, Karlsruhe, FRG).

Animals and Preparations

The experiments were performed in male Wistar rats (Winkelmann, Kirchborchen, FRG) of 150–180 g body weight. The rat was anesthetized by injecting Inactin intraperitoneally (Byk Gulden, Konstanz, FRG) at 120–150 mg/kg body weight. An incision was made to cannulate the urinary bladder with a polyethylene catheter for collecting urine samples. A second incision was made to cannulate the jugular vein to infuse mannitol and test solutions.

The left kidney was exposed by flank incision and by removing the surrounding fascia. For positioning of the kidney, the saddle coil had a slit through which blood vessels and ureter were guided. The kidney was covered with paraffin oil to prevent its drying and to facilitate its placement in the coil. During the MR measurements the body temperature of the animal was maintained at 37°C with a water heating system.

The experiment was started by infusing a mannitol solution (1 ml 20% mannitol, afterwards 3 ml/h 10% mannitol, 5% bovine serum albumin, 4 m*M* KCl, and 2 m*M* $CaCl_2$), which after a few minutes induced osmotic diuresis. Then, the solution containing the ^{19}F marker molecule was administered in a single injection.

During the experiment urine samples were collected in 10- to 15-min periods. The void space delay of the catheter was found to be about 1–2 min. The volume of the urine sample was determined by weighing; the concentration of the ^{19}F-labeled substance was measured by MR spectroscopy.

The probe with the animal was positioned in the magnet's center; the homogeneity of the magnetic field was improved by "shimming" on the proton signals of body water, yielding a linewidth at half height of about 50–60 Hz. Then the ^{19}F MR experiments were started. Due to the broad lines and their low S/N ratio, especially at the end of the experiments, the determination of exact integrals proved unreliable, while the intensity of the signals showed a reduced scatter. Therefore, signal intensity was used to follow concentration changes. This procedure was correct, as the lines showed no variations in their linewidths upon injection of inhibitors for the different transport systems. The calibration of the ^{19}F MR tissue intensities was performed by injecting 0.1 ml solution with a defined concentration of the test substance into the exposed kidney via the kidney vein. Since all other vessels were clamped, signal intensity reflected the amount of injected ^{19}F substance in the kidney.

Results

Mode of Excretion

To describe the renal transport of a test substance, the amounts which are filtered, reabsorbed, and/or secreted must be evaluated. Efficient excretion of our test substances in urine and fast decay of tissue concentration after single intravenous injections made it impossible to collect blood samples regularly. Therefore, the urinary excretion pattern of inulin, which is excreted only by filtration, after bolus injection was obtained from parallel experiments to allow a rough estimate of the fractions of test substance that are filtered, reabsorbed, or secreted. Table 1 exhibits the excreted amounts of TESULF, TFS, TFMH, and inulin determined at similar time periods after injection. Fourteen minutes after injection of 54 µmol TESULF and 6.3 µmol had been excreted (Fig. 1). The excreted fraction of TESULF was calculated by dividing the urinary excreted amount by the injected amount of 54 µmol. The filtration fraction was much smaller with 0.12 than the corresponding value for inulin (0.44). Hence, more than 70% of TESULF as compared to inulin was reabsorbed. For TFMH the corresponding excretion data was 0.41 versus 0.44 for inulin. However, TFMH is bound to plasma proteins to about 60% (Van Ginneken, personal communication), hence, the excretion fraction corrected for protein binding was 0.68, indicating that secretion was about 50% of filtration. For TFS the excretion value was 0.2 against 0.37 for inulin (10 min after injection), indicating 50% reabsorption.

Table 1. Excretion of inulin and ^{19}F test substances at different time intervals after bolus injection

	Time after injection (min)		
	10	14	15
Inulin	0.37	0.44	0.47
TFMH	–	0.41[a] (0.68)	–
TFS	0.20 (Normally fed)	–	0.09 (Starved)
TESULF	–	0.12	–

The amount of a substance excreted in a time interval was divided by the injected amount. The data indicate the excreted fraction of the substance.

[a] 0.68 when taking protein binding into account.

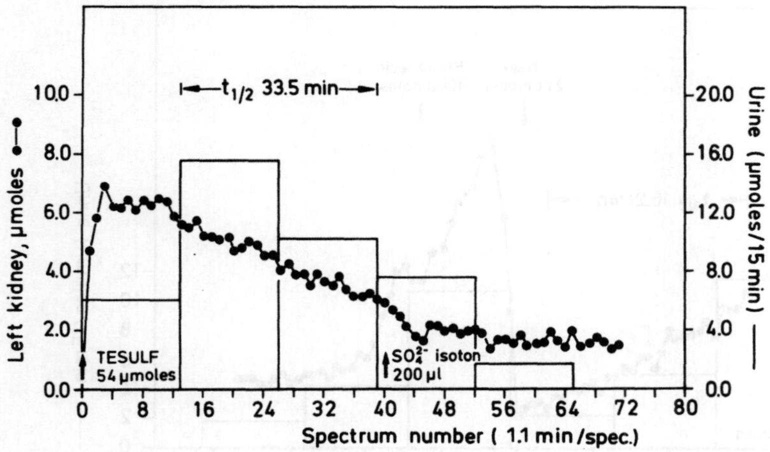

U/T	0.64	1.6	1.4	1.7	0.53
corr.	(0.88)	(2.2)	(1.9)	(2.3)	(0.73)
U/R	0.12	0.39	0.37	0.42	0.13
T/R	0.14	0.25	0.26	0.24	0.24
corr.	(0.11)	(0.19)	(0.20)	(0.18)	(0.18)

Fig. 1. Renal tissue content and urinary excretion of TESULF after intravenous bolus injection of TESULF and sulfate at the indicated times. Number of scans, 64; relaxation delay, 1 s; acquisition time, 0.034 s. Injected volume of TESULF, 1.8 ml; that of sulfate, 1 ml. *U*, Urinary content; *T*, tissue content; *R*, amount of TESULF remaining in the animal (injected amount minus excreted amount); *ratios in parentheses*, derived by correcting tissue content by the amount of TESULF in the lumina of collecting ducts

In starved animals these values change to 0.09 (versus 0.47 for inulin), implying 80% reabsorption.

Excretion Pattern

Figure 2 shows the time dependence of the ^{19}F MR tissue signal intensity due to TFMH after intravenous injection of 27 µmol. The signal reached maximal intensity after 2.2 min and had a decay half time of 16.2 min. After a second dose of TFMH, the PAH transport inhibitor probenecid ($K_{i,PAH}$ 0.03 ± 0.01 m*M*) [9] was injected, inducing a transient increase in the intrarenal TFMH content. This can be explained by liberation of protein-bound TFMH due to competing probenecid binding. Thereafter the intrarenal content of TFMH fell quickly, and the excreted amount of TFMH in urine was also reduced. For quantification of tissue concentration and urinary excretion several ratios were formed using: (a) the kidney tissue content (T)

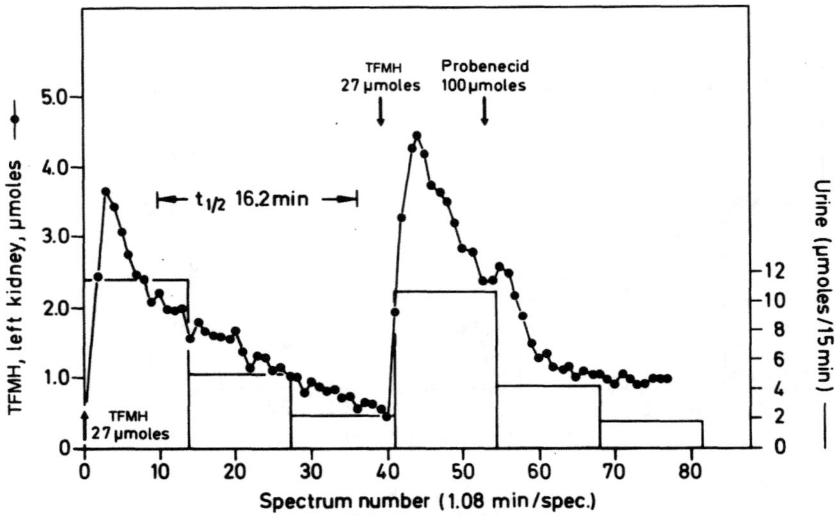

U/T	2.3	1.8	1.5	1.5	1.2	1.2
corr.	(3.2)	(2.5)	(2.1)	(2.1)	(1.7)	(1.7)
U/R	0.55	0.40	0.25	0.37	0.2	0.1
T/R	0.24	0.22	0.16	0.24	0.16	0.1
corr.	(0.17)	(0.16)	(0.11)	(0.17)	(0.11)	(0.07)

Fig. 2. Profiles of renal tissue content and urinary excretion of TFMH after intravenous bolus injections of TFMH and probenecid at the indicated times. Number of accumulated scans, 64; relaxation delay, 1 s; acquisition time, 0.02 s. Injected volume of TFMH, 1.7 ml; that of probenecid, 1 ml. (See Fig. 1 legend)

calculated for the two kidneys, (b) the urinary excreted amount (U), and (c) the amount of substance remaining in the animal (R) for the kidneys. After injection of probenecid the U/T, U/R, and T/R ratios fell. Simultaneous decrease in tissue concentration and decreased urinary excretion of TFMH indicated that the secretion of TFMH was inhibited by probenecid at the contraluminal cell side.

Upon intravenous injection of TFS (Fig. 3a) its ^{19}F MR signal increased steeply. The following decay had a half time of 28.8 min. After the second administration of TFS, succinate was injected, whereby the tissue content of TFS was strongly reduced. Simultaneously the U/T ratio increased transiently from 3.6 to 5.3, indicating that the amount excreted in urine was transiently higher in relation to tissue content. According to our model reabsorption was inhibited by succinate at the luminal cell side.

In an animal that had starved for 3 days, the U/R ratios of TFS (Fig. 3b) were much lower than the corresponding values in normal animals. This

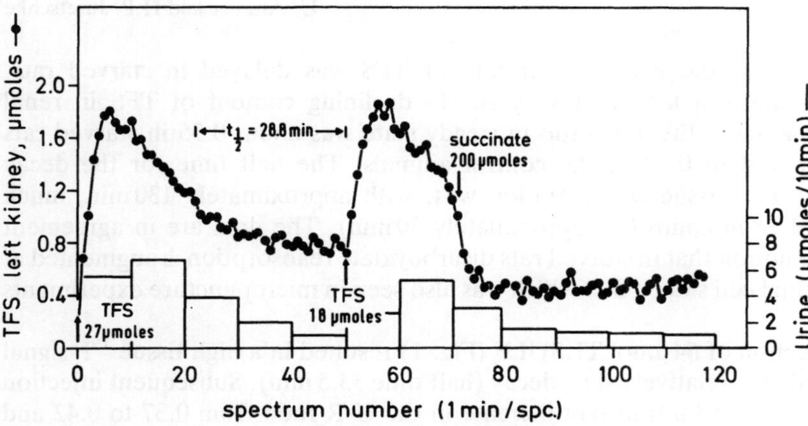

U/T	2.4	3.6	2.9	1.8	0.9	1.3	2.7	4.0	2.6	2.2	2.1	1.6
corr.	(3.2)	(4.8)	(3.9)	(2.4)	(1.2)	(1.7)	(3.6)	(5.3)	(3.5)	(2.9)	(2.8)	(2.1)
U/R	0.33	0.56	0.45	0.30	0.18	0.17	0.38	0.29	0.17	0.15	0.17	0.15
T/R	0.14	0.15	0.15	0.16	0.18	0.13	0.14	0.07	0.06	0.07	0.08	0.10
corr.	(0.11)	(0.11)	(0.11)	(0.12)	(0.14)	(0.10)	(0.11)	(0.05)	(0.05)	(0.05)	(0.06)	(0.08)

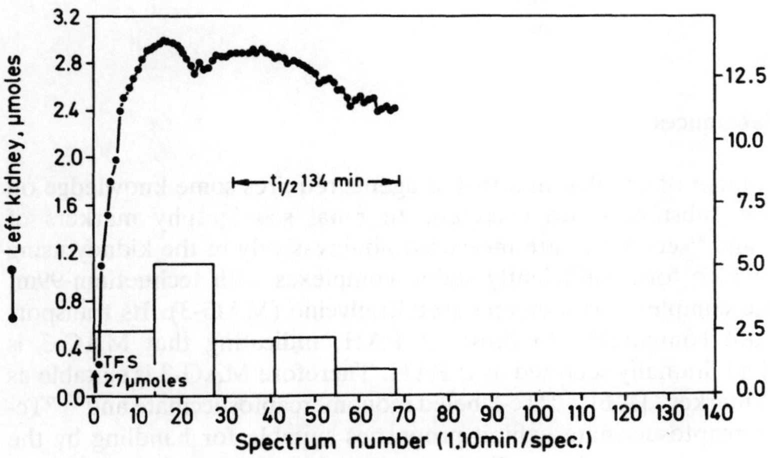

U/T	0.57	0.89	0.37	0.42	0.18
corr.	(0.63)	(0.98)	(0.41)	(0.46)	(0.20)
U/R	0.10	0.23	0.12	0.14	0.07
T/R	0.17	0.26	0.32	0.33	0.39
corr.	(0.14)	(0.21)	(0.26)	(0.27)	(0.32)

Fig. 3. *Top*, Renal tissue content and urinary excretion of TFS after intravenous bolus injections of TFS and succinate at the indicated times. Number of scans, 240; relaxation delay, 0.24 s; acquisition time, 0.1 s. Injected volume of TFS, 0.45 ml; that of succinate; 1 ml. *Bottom*, Renal tissue content and urinary excretion of TFS in a rat after 3 days of starvation. TFS was injected intravenously in a single dose. Number of scans, 64; relaxation delay, 1 s; acquisition time, 0.03 s. Injected volume of TFS, 0.3 ml. (See Fig. 1 legend)

indicates that the urinary excretion of TFS was delayed in starved rats. Concomitantly a high and very slowly declining content of TFS in renal tissue was seen: the T/R ratio in steady state was about 0.35 in starved rats as compared to 0.13 in the control animals. The half time for the decay of renal TFS tissue concentration was, with approximately 130 min, much longer than in controls (approximately 30 min). The data are in agreement with the notion that in starved rats dicarboyxlate reabsorption is augmented at the luminal cell side, a fact which was also seen in micropuncture experiments [8].

Injection of 54 μmol TESULF (Fig. 1) resulted in a high tissue ^{19}F signal followed by a relatively slow decay (half time 33.5 min). Subsequent injection of sulfate caused a transient increase in the U/R ratio from 0.37 to 0.42 and in the U/T ratio from 1.9 to 2.3. This indicates that the urinary excretion of TESULF was augmented by the injection of sulfate, while tissue concentration decreased (however, only transiently so that it did not show up in the T/R ratio). This pattern indicates an inhibition of TESULF reabsorption at the luminal cell side.

Discussion

Choice of Substances

The development of suitable fluorinated agents requires some knowledge on how the test substances are excreted. In renal scintigraphy markers of "filtration" and "secretion" are measured noninvasively in the kidney using substances which form sufficiently stable complexes with technetium-99m. One of these complexes is mercaptoacetyltriglycine (MAG-3). Its transport properties are comparable to those of PAH, indicating that MAG-3 is filtered and additionally secreted as is PAH. Therefore MAG-3 is suitable as a "secretion marker" [1, 6]. ^{99m}Tc-labeled monomercaptosuccinate and ^{99m}Tc-labeled dimercaptosuccinate exhibit properties suitable for handling by the dicarboxylate transport systems. Both compounds are filtered and taken up by the cells of the proximal tubule. The overall reabsorption is enhanced during starvation [4, 5]. ^{99m}Tc-labeled diethylenetriaminopentaacetic acid (DTPA) is used as marker of "filtration" [2]. This substance forms stable complexes and has no affinity for the PAH and dicarboxylate transport system.

With these substances used in renal scintigraphy, we searched for comparable ^{19}F-containing compounds which are filtered and additionally secreted or reabsorbed. The required substances should have a high specifity for one of the transport systems for anions described for the proximal tubule: (a) for the transport system of hydrophobic organic anions (PAH transport system) in the contraluminal membrane; (b) for the Na^+-coupled transport systems for dicarboxylates in the luminal and contraluminal membrane; (c) for the sulfate/oxalate transport system that is Na^+ dependent (luminal membrane) or Na^+ independent (contraluminal membrane). For the *PAH transport system* TFMH was synthesized ($K_{i,PAH}$ 0.03 mM). About 60% of TFMH is bound to plasma proteins; Van Ginneken observed metabolization of TFMH to 3-trifluoromethylbenzoic acid in dogs, which, however, was detectable only in plasma, not in urinary samples (personal communication). Also in our experiments no metabolites were detected in urine of the rat. Hence we believe that metabolization has no impact on our studies. The only restriction for the use of TFMH is its low water solubility. For the *dicarboxylate transport system* TFS was prepared ($K_{i,succ}$ 0.05 mM). TFS also possesses a small affinity for the PAH transport system ($K_{i,PAH}$ 1.4 mM). TFS was not metabolized. For the *sulfate transport system* TESULF ($K_{i,sulf,cl}$ NS; $K_{i,sulf,l}$ 65 mM) was synthesized because, according to the known substrate specificity of the sulfate transport system, we expected, an interaction. The measured affinity for the luminal sulfate transport system was low. For the contraluminal transport system an interaction with TESULF was not demonstrated at all. Unfortunately, it has not been possible up to now to find a ^{19}F-labeled substrate with higher affinity for the sulfate/oxalate transport system.

For TFS about 50% of the filtered amount was reabsorbed. Reabsorption increased in starved animals to about 80%. TESULF was also reabsorbed by 70%, while TFMH was secreted by about 50%, i.e., the urinary excreted amount was twice that of the filtered amount.

Results of In Vivo ^{19}F MR Experiments and Applications in Pharmacology

In vivo ^{19}F MR spectroscopy allows continuous monitoring of suitable test substances in kidney tissue and correlation to the urinary excretion. We demonstrated that the hippurate analog TFMH is secreted. Secretion was inhibited by probenecid, which causes a marked decrease in both urinary and tissue concentrations of TFMH. This indicates that the side of interaction is the contraluminal cell membrane. The increase in urinary TFS excretion after succinate injection was accompanied by a decreased tissue TFS content, suggesting succinate-inhibited TFS absorption through the membrane. The substrate for the sulfate transport system, TESULF, was also reabsorbed in spite of its low affinity for the transporter. Injection of sulfate caused, similar to TFS, an increase of the urinary excretion of TESULF and a decrease in tissue content. The reabsorption of TESULF was therefore inhibited by sulfate at the luminal cell side.

Quantification of Tissue Content

The method used (injecting a defined amount of the test substance into the kidney, clamping all vessels, and measuring the MR signal under comparable conditions) was expected to yield a reliable calibration of tissue content since the distribution of the substance in the kidney should be similar to that during the in vivo MR experiment.

The detected tissue signal consists of contributions from different compartments of the kidney. The total renal volume contains about 13.07% blood capillaries, 48% epithelium (33% proximal tubular cells), 26.8% lumina (15.3% collecting duct), and 10.3% extracellular space [7]. During the experiment the compartments contained different concentrations of the test substance. The urine present within the renal tubules, especially within the lumina of the collecting ducts, must be considered because high concentrations of a test substance can be reached. We tested two methods. Administration of MR shift reagents was expected to cause distinct signals from the intra- and extracellularly located ^{19}F-labeled test substance. In in vitro experiments the influence of Tb, Ho, Dy, Er, and Gd chelated with ethylenediaminotetraacetic acid (EDTA), DTPA, and in the case of Dy with triethylenetetraminohexaacetic acid on the chemical shift of the resonance signal of TFMH was investigated, but the detected shifts were too small (e.g., 2.6 ppm by 14 m*M* Ho-EDTA). Administration in in vivo experiments resulted in a strong increase in linewidth up to the disappearance of the signal. In these experiments it was not possible to distinguish between intra- and extracellular signals. The line broadening was probably caused by the inhomogeneous distribution of the shift reagent in the kidney.

Furthermore, we evaluated the content of the test substance in the lumina of the tubules by measuring the rate of excretion and the contact time in the kidney between appearance of the test substance in the blood capillaries and their appearance at the tip of the renal papilla. For this purpose we used bolus injections of lissaminegreen, which is mainly filtered. The contribution of the urinary content to total tissue signal was estimated from these dye studies to be about 25%. The necessary correction of the detected tissue signal therefore depends on the excretion of a test substance and the resulting urinary content during the MR measurement. The urinary content is higher when the substance is filtered and additionally secreted compared to a substance which is filtered and reabsorbed. For TFMH, which is filtered and secreted, the estimated correction for the measured tissue signal is in the range of 27%. Accordingly all ratios calculated in Fig. 2 had to be corrected in relation to the tissue content; the corrected ratios are shown in parentheses. Tissue signals from TFS (Fig. 3a), which is filtered and reabsorbed, were corrected by a urinary content of about 25%. In the starved rat, where the reabsorption is increased, the correction was in the range of 16% because the urinary excretion of TFS is reduced (Fig. 3b).

Finally, for TESULF (Fig. 1) the correction of the tissue signal for urinary content was 27%.

The applied method is suitable in numerous investigations where the measurement of physiological or pathophysiological changes in tissue concentrations and their relation to changes in their urinary excretion are important. The described nontoxic ^{19}F-labeled substances are useful if they can be detected in patients by ^{19}F MR spectroscopy. A replacement of ^{99m}Tc-containing complexes by administration of ^{19}F-containing test substances is therefore conceivable.

The study of kidney function in vivo is possible for the detection of pathological events or toxic interactions of any compound. Obviously, studies with compounds interacting with the dicarboxylate transport systems must be interpreted with great caution since they are taken up into the cells from both cell sides. Very often these substances reach high intracellular levels, which are further increased during starvation when the tubulary reabsorption is increased. In this situation high intracellular concentrations could cause toxic effects. Administration of in vivo ^{19}F MR spectroscopy allows one to monitor the intracellular content of a substance and to find compounds which may prevent or avoid a toxic accumulation.

Summary

Using in vivo^{19}F MR spectroscopy it is possible to monitor the content of ^{19}F-labeled marker molecules in the rat kidney and its urine. Transport of the markers can be characterized by administration of specific inhibitors. TFMH is suitable as marker for secretion. Inhibition of TFMH secretion takes place at the contraluminal cell side. TFS and TESULF are suitable markers for reabsorption. Inhibition of reabsorption by interfering succinate and sulfate, respectively, occurs at the luminal cell side. These studies provide a basis for the replacement of ^{99m}Tc-containing test substances by nontoxic ^{19}F reagents in renal diagnostics.

Acknowledgements. We thank Prof. K.J. Ullrich (MPI für Biophysik, Frankfurt) for valuable discussions and helpful ideas concerning the work, Dr. F. Fahrenholz and his group (MPI für Biophysik, Frankfurt) for the synthesis of TFMH, and Prof. Dr. H. Fasold (Institut für Biochemie, Universität Frankfurt) for the synthesis of TESULF. We thank Mr. G. Rumrich for an introduction to animal preparation techniques and Prof. Dr. H. Rüterjans (Institute für Biophysikalische Chemie, Universität Frankfurt) for his support and discussions.

References

1. Bubeck B, Brandau W, Seinbächer M, Reinbold F, Dreikorn K, Eisenhut M, Georgi P (1988) Technetium-99m labelled renal function and imaging agents: II. Clinical evaluation of Tc-99m MAG-3. Nucl Med Biol 15/1:109–118
2. Clifton GC, Anderson C, McMahon G, Vargas R, Wallin JD (1989) Monoexponential analysis of plasma disappearance of Tc-99m DTPA and I-131 iodohippurate: a reliable method for measuring changes for renal function. J Clin Pharmacol 29/5:466–71
3. Fritzsch G, Haase W, Rumrich G, Fasold H, Ullrich KJ (1984) A stopped flow capillary perfusion method to evaluate contraluminal transport parameters of methylsuccinate from interstitium into renal proximal tubular cells. Pflugers Arch 400:250–256
4. Kamper AL, Thomsen HS, Nielsen SL, Strandgaard S (1990) Renal uptake of dimercaptosuccinic acid and glomerular filtration rate in chronic nephropathy at angiotensin converting enzyme inhibition. Eur J Nucl Med 16:85–88
5. Lauer O, Heidenreich P, Reidel G, Hör G, Pabst HW (1980) Clinical experience with Tc-99m MMSA, a new renal imaging agent, as compared with Tc-99m DMSA. In: Hollenberg NK, Lange S (eds) Radionuclides in nephrology. Thieme, Stuttgart, pp 23–31
6. Müller-Suur R, Müller-Suur C (1989) Glomerular filtration and tubular secretion of MAG-3 in the rat kidney. J Nucl Med 30:1986–1991
7. Pfaller W (1982) Structure function correlation on rat kidney. Adv Anat Embryol Cell Biol 70:22–25
8. Ullrich KJ, Fasold H, Rumrich G, Klöss S (1984) Secretion and contraluminal uptake of dicarboxylic acids in the proximal convolution of rat kidney. Pflugers Arch 400:241–249
9. Ullrich KJ, Rumrich G (1990) The frontiers of nephrology. In: Berliner RW, Honda N, Ullrich KJ (eds) International Congress Series 877. Excerpta Medica, Amsterdam, pp 55–66

Magnetic Resonance Evaluation of Experimental Renal Vein Thrombosis

D. Uhlenbrock[1]

Acute cases of renal vein thrombosis occur primarily among children; in the case of adults, however, a gradual subacute development of the disease can be observed. Renal vein thrombosis is often an accompanying disease in the case of a retroperitoneal process, such as absess, pancreatitis, retroperitoneal tumor, or trauma. It can also be found in cases of thromboembolic disease and depletion of the extracellular volume (especially in the case of children), as well as when there are various kinds of nephropathy, such as amyloidosis, diabetes mellitus, polyarteritis nodosa, lupus erythematosus, and membranous diseases.

The symptoms become clinically manifest in the form of flank pain, temperature, nausea, vomiting, and potentially hypertension [1, 5, 6]. Generalized edema occurs. Oliguria or renal failure is possible. The clinical relevance of this disease induced us to investigate the ability to determine renal vein thrombosis experimentally by means of magnetic resonance.

Material and Methods

For the purpose of this study, 25 Han-Wistar rats were examined. In terms of its anatomical construction, the kidney of the rat is somewhat different from that of the human. It has only one papilla ending in one large calix (Fig. 1). By means of a transperitoneal incision, access to the left kidney of the rats was chosen. The adipose capsula was completely removed in all cases. The renal vein was prepared and ligated close to the hilum of the kidney. A short time after the vein ligature a marked swelling of the kidney occurred, with a dark red-brown coloring appearing on its surface. This can be seen as a criterion for a successful ligature. In addition to the preoperative measurement of the animal, further measurements were taken 2, 8, and 30 h after the ligature and 2–4, 6, 7, 14, and 21 days after the ligature.

Measurements were taken in spin-echo technique using T1-weighted sequences (TR/TE 500/30), proton-weighted sequences (TR/TE 1500/30),

[1] Department of Radiology, St. Vincenz-Krankenhaus, Am Busdorf 2-4a, D-33098 Paderborn, FRG

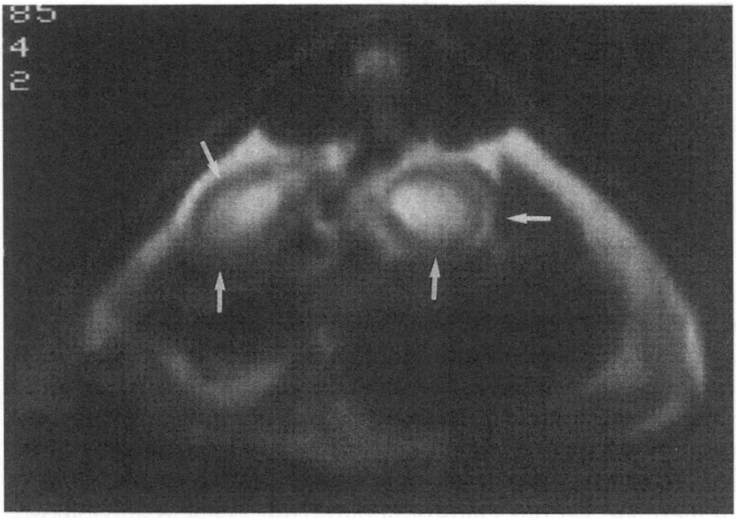

Fig. 1. Transverse cut of kidneys of a healthy rat before operation (T2-weighted sequence). Left and right kidneys with typical appearance. The kidney of the rat has only one papilla, ending in one large calix. *White arrows*, left and right kidneys

and multiecho sequences (TR/TE 1500, 30, 60, 90, 120, 150, 180, 210, 240). The matrix was 256 × 256; slice thickness was 5 mm. Based on the sequences with the different TR times, the T1 relaxation times were calculated. The T2 relaxation times of the kidney were measured, based on the multiecho sequences. In the following study, however, only the T2 relaxation times are explained further. Due to their rather large standard deviation, the T1 relaxation times were not usable; further improvements in the software are needed here to obtain usable data. During the 21 days, 15 of the 25 animals were killed to collect histological data. Thus, measurements could still be taken on ten animals 21 days after the ligature.

Results

The T2 relaxation times for the renal cortex taken prior to the operation were 75.8 ms on the right and 77.33 ms on the left. For the renal medulla the preoperative measurements were 122.95 ms on the right and 121.87 ms on the left. After the vein ligature on the left side there was a constant increase in the relaxation time of the renal cortex during the first 30 h, reaching 97 ms on the left side. On the right there was no significant change in the data (Fig. 2). The increase in relaxation time for the renal cortex on the left was highly significant ($p < 0.001$). During the following examinations there was a constant decrease in T2 relaxation time for the left side in the area of the

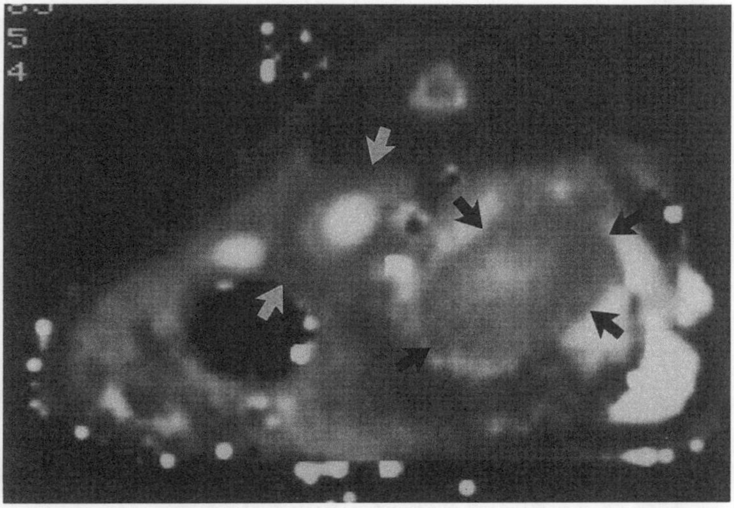

Fig. 2. T2-calculated image 8 h after the renal vein ligature on the left. Increase in relaxation time of the left renal cortex (left renal cortex 82 ms, right side 77 ms) and decrease in relaxation time of the left renal medulla (left side 100 ms, right side 124 ms). Note the marked increase in size of the left renal cortex. *Black arrows*, left kidney; *white arrows*, right kidney

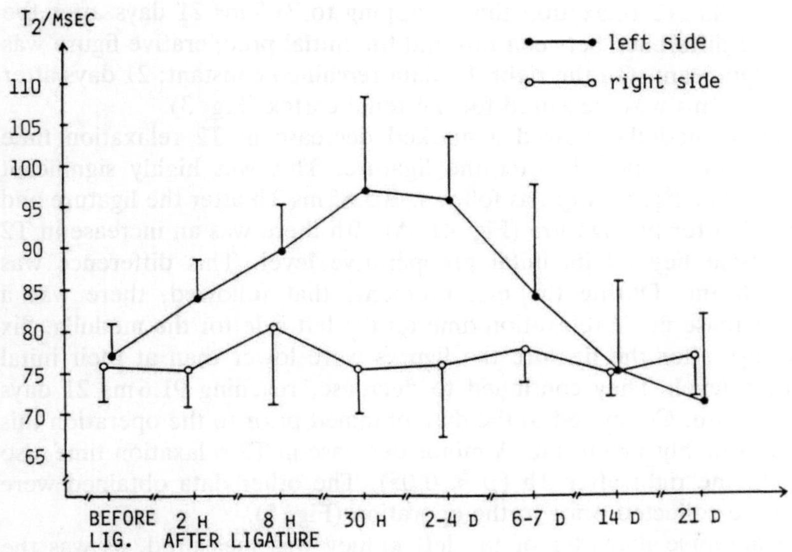

Fig. 3. T2 relaxation times of the renal cortex before and after ligature of the left renal vein

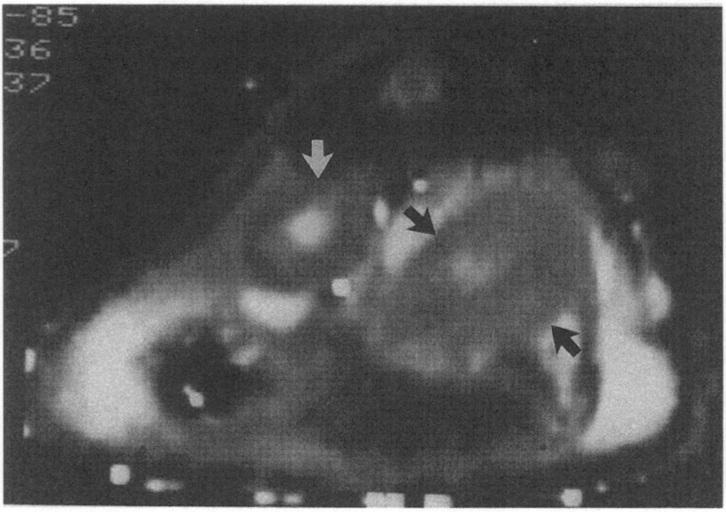

Fig. 4. T2-calculated image. Examination 8 h after the ligature of the left renal vein. Marked decrease in T2 relaxation time of the left medulla, compared with the right side and the preoperative values (left 100 ms, right 122 ms). Increase in T2 relaxation time of the left cortex (left 88 ms, right 78 ms). Swelling of the left kidney, based only on the enlarged cortex. *Black arrows*, left kidney; *white arrows*, right kidney

renal cortex, the T2 relaxation time dropping to 71.7 ms 21 days after the ligature. The difference between this and the initial preoperative figure was no longer significant. On the right the data remained constant; 21 days after the ligature, 77 ms was measured for the renal cortex (Fig. 3).

The renal medulla showed a marked decrease in T2 relaxation time on the left side 2 and 8 h after the ligature. This was highly significant ($p < 0.001$). The figures were as follows: 105.85 ms 2 h after the ligature and 106.89 ms 8 h after the ligature (Fig. 4). At 30 h there was an increase in T2 relaxation time beyond its inital preoperative level. This difference was highly significant. During the measurements that followed, there was a constant decrease in T2 relaxation time on the left side for the medulla. Six to seven days after the ligature the figures were lower than at their inital preoperative level. They continued to decrease, reaching 91.6 ms 21 days after the ligature. Compared to the data obtained prior to the operation this decrease was highly significant. A minor decrease in T2 relaxation time also occurred on the right after 2 h ($p < 0.05$). The other data obtained were close to those collected prior to the operation (Fig. 5).

The transverse diameter of the left kidney was measured, as was the diameter of the cortex and the medulla, each separately. The total diameter of the kidney increased constantly, reaching a maximum 30 h after the ligature (Fig. 6). This increase was highly significant. In the course of the study, there was a constant decrease in the size of the kidney (Fig. 7); 21

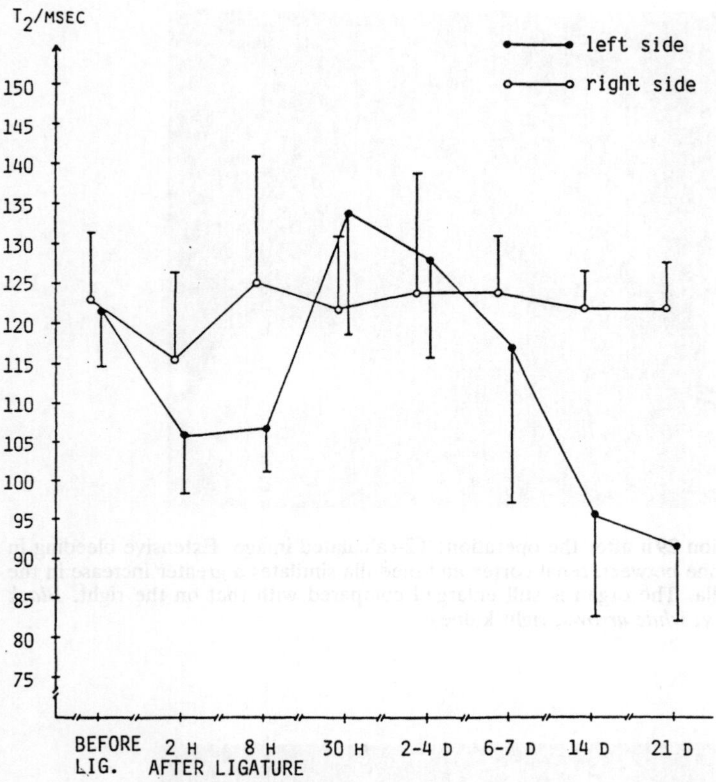

Fig. 5. T2 relaxation times of the renal medulla before and after ligature of the left renal vein

days after the ligature the values were clearly lower than the initial preoperative results. The right kidney showed no significant change postoperatively during the first hours and days after the ligature when compared to the data obtained preoperatively. The measurements taken 6 and 7 days after the ligature showed a slight increase in size ($p < 0.05$). Another marked increase in size occurred later, the data having changed significantly compared to those obtained prior to the operation.

The separate observation of the cortex and the medulla showed a highly significant increase in size on the left side of the cortex as early as 2 h after the ligature. The maximum value was measured 2–4 days after the ligature. Later the diameter of the left renal cortex decreased in diameter to a point below that of its initial size (Fig. 8). The renal medulla did not show a very significant increase in size until 30 h after the ligature. Later, however, there was a decrease in the size of the renal medulla to a point below its inital value, the difference being highly significant 21 days after the ligature (Figs. 9, 10).

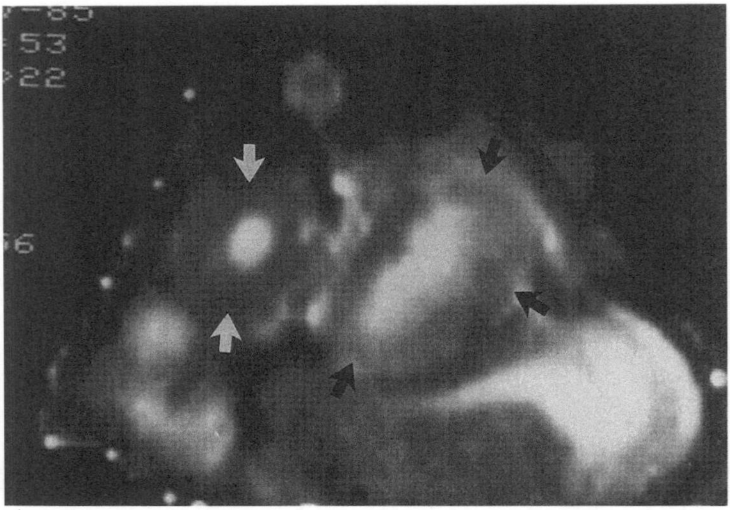

Fig. 6. Examination 29 h after the operation; T2-calculated image. Extensive bleeding in the transitional zone between renal cortex and medulla similates a greater increase in the size of the medulla. The organ is still enlarged compared with that on the right. *Black arrows*, left kidney; *white arrows*, right kidney

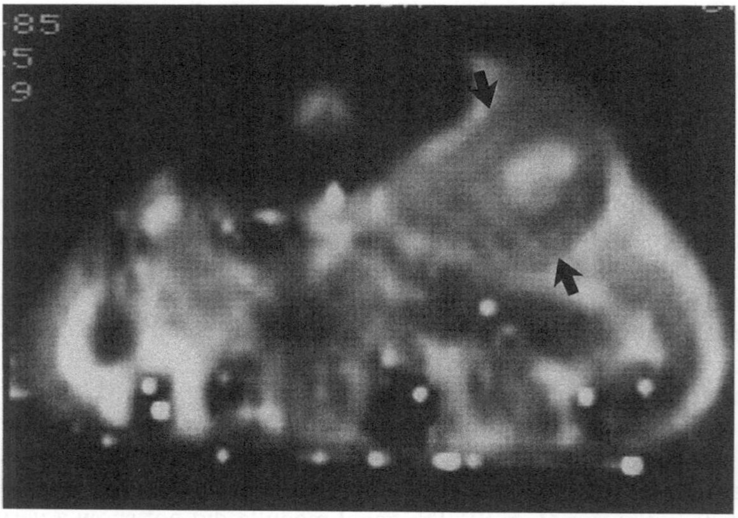

Fig. 7. Examination 3 days after ligature. Decrease in size of the kidney compared with the images 2 and 8 h after the ligature (not shown), but the organ is still enlarged. *Black arrows*, left kidney

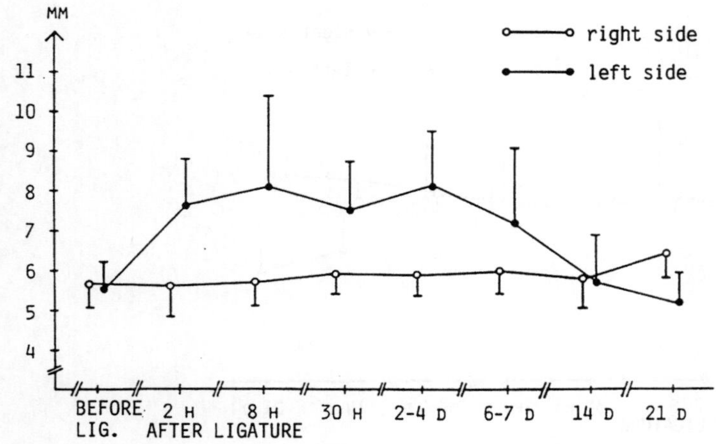

Fig. 8. Diameter of renal cortex before and after ligature of the left renal vein

From a histological point of view, hemorrhages appeared only a short time after the vein ligature in the area of the medulla and in the transitional zone between the medulla and the cortex. In connection with the vasconges-tion there was hemorrhaging of the hilar adipose tissue. Macroscopically there was a considerable enlargement of the organ as a result of these changes. Necrosis first appeared on the proximal tubular epithelium whereas glomeruli and the distal tubule structures, studied by light microscopy, could still be seen as being intact 8 h after the ligature. Fifteen hours after the ligature there was complete necrosis of the tubules, and at 30 h the glomeruli had also been affected.

About 3–5 days after the ligature the bleeding had been largely resorbed, the remains of which were found in the form of siderophages. The necrosis was striking in this respect as it was anemic at this point, but the kidney was still enlarged compared to the right side. As the necrosis progressed, the size decreased. After 21 days the affected kidney was smaller than the other.

It was striking that the necrotizing in one organ did not always proceed uniformly, but rather there were zones of complete necrosis and zones of partial necrosis which were found next to one another. Twenty-one days after the ligature isolated foci of tissue calcification were found, as well as the beginning retrograde epithelization of the tubules by urothel cells.

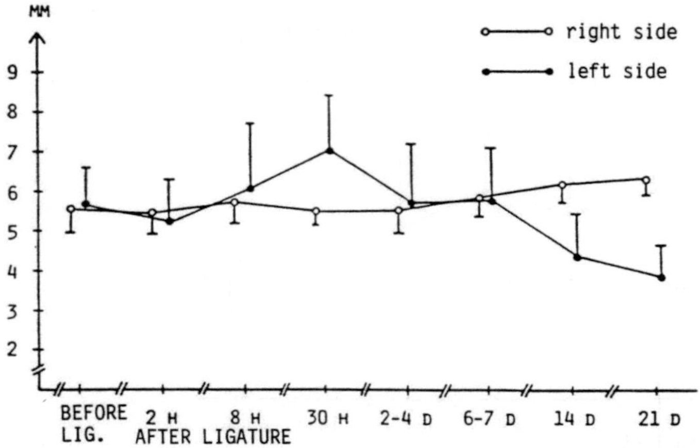

Fig. 9. Diameter of renal medulla before and after ligature of the left renal vein

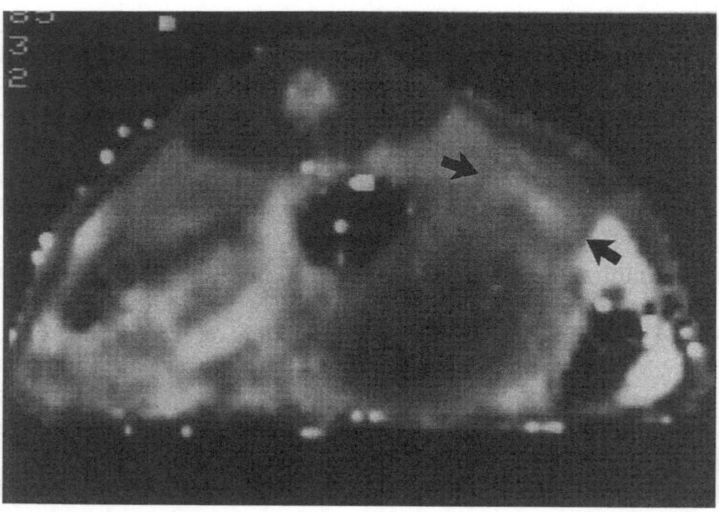

Fig. 10. Contracted kidney 20 days after the ligature. T2-calculated image, which only shows the left kidney (left renal cortex 74 ms, on the right 77 ms, left medulla 96 ms, on the right 128 ms). *Black arrows*, left kidney

Discussion

The studies on renal vein occlusion published to date cannot be compared with the results presented in this paper [4, 7–9]. The experimental results presented here provide a more differentiated view than was thought possible, based on the data published to date. Especially with examination of the development of the disease, the changes could be determined more exactly than those published earlier. In the course of a few days very contrary processes occurred, as can be seen in the initial enlargement of the kidney following the operation and its shrinking as a result of necrosis. Something similar could be observed in the case of T2 relaxation time, which at first showed an increase in the area of the cortex. After about 6 days, however, it decreased; the values were clearly lower 3 weeks after the ligature than the control values. In the area of the medulla, exactly the opposite occurred.

Another striking feature to note is that the enlargement of the kidney in the first few days was solely the result of an increase in the volume of the cortex. The medulla, on the other hand, showed no change. During the 2nd and 3rd weeks after the vein ligature a decrease in the diameter of the medulla occurred. The enlargement of the cortex might have been due to an edema and a stasis in the arteries and veins, in connection with which the inner cortex zone was primarily affected. The findings of a selective swelling of the cortex were dealt with similarly in a study by Hricak and colleagues [3].

The reduction of T2 relaxation time for the renal medulla depended primarily on a decrease in urinary discharge accompanied by a reduced collection of urine in the tubules and in the collecting tubules. The increase in T2 relaxation time in the medulla, occurring about 8 h after the ligature, probably resulted from necrosis of the epithelial cells of the tubules, situated primarily in the outer medulla. At this time, the necrosis had become clearly manifest [2]. This cell necrosis occurred along with an edema of the cells, that is, with the interstitial tissue as well. Necrosis also occurred in the glomeruli in the area of the cortex. Here too an edema could have been the consequence. However, it does not lead to an increase in T2 relaxation time to the same extent as in the medulla. Rather, a reduction in relaxation time for the cortex developed as a result of the decrease in congestion and of the beginning resorption of the interstitial bleedings.

Summarizing the results, we can say that the existence of a contracted kidney means a reduction in T2 relaxation time for the medulla. For the cortex, however, T2 relaxation time remains largely unchanged (Fig. 10). Hence it follows that the typical description of the contracted kidney on T1-weighted images is characterized by the fact that the contrast between cortex and medulla is reduced, especially as a result of an increased signal in the area of the medulla because of a loss of free water in the tubules.

It remains to be seen whether the good results obtained in the experiment can be applied clinically, and to what extent diagnostic guidelines can be developed from these findings and used when there is suspicion of renal vein thrombosis.

References

1. Brass H, Deppermann D (1986) Das nephrotische Syndrom. Diagnostik und Therapie. Therapiewoche 36:344–353
2. David H, Uerlings I (1967) Elektronenmikroskopische Untersuchungen zur Entdifferenzierung und Regeneration des Tubulusepithels der Niere bei chronischer Nierenvenendrosselung. Exp Pathol 1:273–290
3. Hricak H, Sandler MA, Madrazo BL, Eyler WR, Sy GS (1981) Sonographic manifestations of acute renal vein thrombosis: an experimental study. Invest Radiol 16:30–35
4. London DA, Davis PL, Williams RD, Crooks LE, Sheldon PE, Gooding CA (1983) Nuclear magnetic resonance imaging of induced renal lesions. Radiology 148:167–172
5. McCarthy LJ, Titus JL, Daugherty GW (1963) Bilateral renal vein thrombosis and the nephrotic syndrome in adults. Ann Intern Med 58:837–857
6. Panner B (1963) Nephrotic syndrome in renal vein thrombosis. Arch Pathol 76:303–317
7. Slutsky RA, Andre MP, Mattrey RF, Brahme FJ (1984) In vitro magnetic relaxation times of the ischemic and reperfused rabbit kidney: concise communication. J Nucl Med 25:38–41
8. Winfield AC, Gerlock AJ, Shaff MI (1981) Perirenal cobwebs: a CT sign of renal vein thrombosis. J Comput Assist Tomogr 5:705–708
9. Yuasa Y, Kundel HL (1985) Magnetic resonance imaging following unilateral occlusion of the renal circulation in rabbits. Radiology 154:151–156

Magnetic Resonance and Organ Viability

The Role of Magnetic Resonance in Cryopreservation of Tissues

D.E. Pegg[1]

Introduction

The preservation of viable cells at very low temperatures became a reality in 1949 when Polge and colleagues accidentally discovered that the incorporation of glycerol into the solution in which fowl spermatozoa were frozen enabled them to survive exposure to a temperature of −80°C [8]. They correctly realised that cells that could survive such an experience could probably be stored for very long periods of time at −80°C since this temperature would effectively arrest all chemical processes. Their experiment provided the foundation for the cryopreservation of cattle sperm, which is crucial to modern animal husbandry. Lovelock, working in the same institute, was quick to provide the basic explanation for the cryoprotective effect of glycerol, which was that it reduces the amount of ice that is formed at any given temperature, and that because it is itself remarkably non-toxic, is able to moderate the otherwise lethal effect of freezing [5]. It had been supposed until that time that ice has a directly damaging effect on the cells, but Lovelock advanced an alternative, and convincing, argument that the effect is indirect; that the formation of ice causes the dissolved solutes to increase in concentration in the remaining liquid phase; and that it is this concentration, particularly of salts, that damages the cells. Subsequent studies by Mazur [6] revealed that intracellular freezing is indeed directly damaging, but that control of the cooling rate provides a simple means of ensuring that all the ice is extracellular, where it has little or no effect upon most cell suspensions. This, very briefly, is the basis of all the currently used techniques of cell cryopreservation, which have found many important applications in medicine and agriculture.

In the field of medicine and surgery, it was natural to attempt to use the same techniques to cryopreserve tissues and ultimately whole organs, which are used in various therapeutic procedures. Limited success has been achieved, notably with heart valves and skin, but in many other

[1] MRC Medical Cryobiology Group, University Department of Surgery, Douglas House, Trumpington Road, Cambridge CB2 2AH, UK

cases the results have been unsatisfactory, and the frozen and thawed tissues unusable. In particular, it has proved quite impossible to develop a technique for the cryopreservation of whole transplantable organs such as kidneys. It has now become clear that the principal cause of the difficulty is a direct consequence of the formation of ice outside the cells, but inside the tissue or organ that is to be preserved. Although the cells may survive in large numbers, the ice forming within the extracellular matrix disrupts the tissue and renders it functionless [7].

Thus, it has become clear that there are two approaches available for the preservation of tissues. The first is to use temperatures above the freezing point, in which case ice is avoided, but the temperature is sufficiently high for metabolism to continue. The second approach is to use much lower temperatures, but in the presence of extremely high concentrations of cryoprotective solutes so that either no ice is formed, or the amount of ice that is formed is insufficient to produce lethal injury. Magnetic resonance (MR) can assist in both approaches. In the first, phosphorus MR can be used to monitor the energy status of the still-metabolising tissue. In the second case, it can be used to monitor the transport of cryoprotectants into the organs or tissues, such that time-efficient schedules of addition and removal can be used and toxicity thereby reduced. For the future, there is also the possibility that MR microscopy could be used to locate and quantify ice in partially frozen tissue and organs.

Measurement of Energy Status of Cells in Tissues and Organs

Many studies have been published showing the loss of high energy compounds from tissues and organs during ischaemia and storage. For example, in one such study on the rabbit kidney, it was shown that the concentration of ATP and ADP falls during warm ischaemia (ischaemia at 37°C) to very low levels within 15 min. It is clear that the maintenance of supplies of stored chemical energy will be crucial to the eventual restoration of function. In tissue and organ preservation for subsequent transplantation, the measurement of energy status has two potential functions: first, to compare several methods of preservation to determine which is most likely to be effective; second, to assess the status of an individual organ or tissue – this is sometimes called "viability testing". For the first purpose, the ability of phosphorus MR to make non-invasive measurements is an advantage, and for the second it is essential. Figure 1 shows spectra from a study in which a standard cardioplegic solution (St. Thomas' Hospital solution) was compared with a specially designed preservation solution (CP5) for the storage of rabbit hearts at 10°C [4]. The hearts were perfused with the relevant solution, stored at 4°C and placed in the spectrometer at various time intervals during storage. The peaks characteristic of creatine phosphate

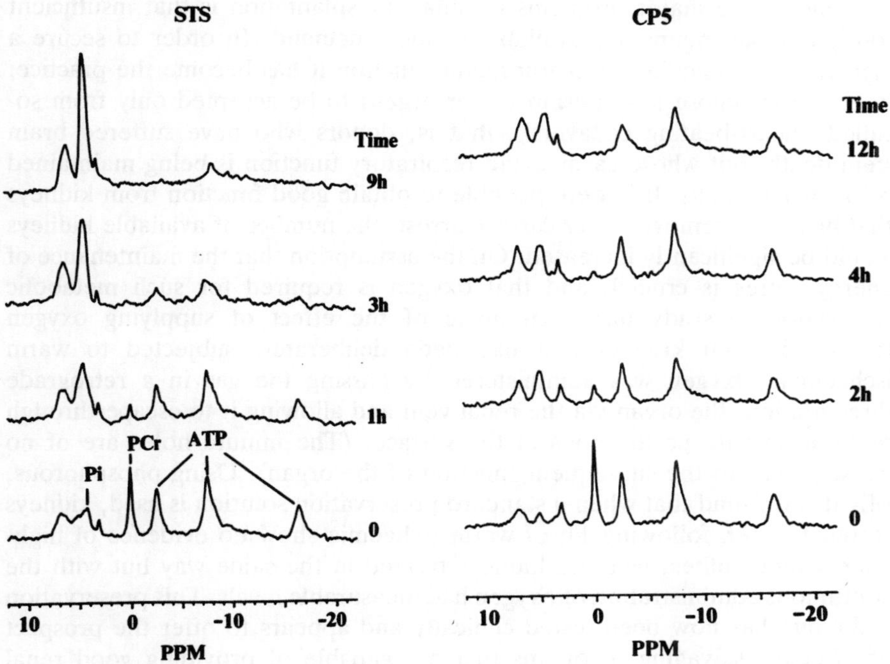

Fig. 1. ^{31}P MR spectra of two hearts, one preserved with St. Thomas' Hospital solution (STS) and the other with CP5. Chemical shifts are assigned in parts per million (ppm) relative to the phosphocreatine (*PCr*) signal, which is assigned the value 0 ppm. Signals are assigned (*from right to left*) to the β, α and γ phosphates of ATP, phosphocreatine (*PCr*), and inorganic phosphate (*Pi*). The peaks at 3 ppm and 6–7 ppm are from phosphodiesters and phosphomonoesters, respectively. The ATP peak at −16 ppm was used in the data analysis, as the two signals at −2.5 and −7.5 ppm contain possible contributions from ADP and, in the case of the −7.5 ppm peak, from NAD and NADH. The times alongside each spectrum represent the time after first placing the heart in the magnet, which was typically 2–5 min after flushing. Each spectrum was accumulated using 64 pulses at intervals of 1 s, and a line-broadening function of 25 Hz was applied prior to Fourier transformation. (From [4] with permission)

and ATP remained detectable longer with the new solution than they did with the old cardioplegic solution, and the concomitant increase in the inorganic phosphate peak was more modest. In fact, measurement of the ratio of the ATP peak to the total peak area showed total discrimination between the two preservation solutions from 4 h of storage onwards. This is an example of the use of phosphorus MR to select between two solutions. Phosphorous MR has also been used as a screening test for human kidneys that are to be transplanted. When seven kidneys that were transplanted were ranked by post-transplant function, and this ranking was compared with measurements of high-energy phosphates by MR made prior to grafting, substantial agreement was reported [3].

One of the major problems in renal transplantation is that insufficient transplantable organs are available to meet demand. In order to secure a high rate of immediate post-transplant function it has become the practice, in countries whose laws permit it, for organs to be accepted only from so-called "heart-beating cadavers", that is, donors who have suffered brain stem death, but whose cardiac and respiratory function is being maintained by artificial means. If it were possible to obtain good function from kidneys that had been removed *after* cardiac arrest, the number of available kidneys would be significantly increased. On the assumption that the maintenance of energy stores is crucial, and that oxygen is required for such metabolic restoration, a study has been made of the effect of supplying oxygen to experimental kidneys that had been deliberately subjected to warm ischaemia. Oxygen was administered by passing the gas in a retrograde direction into the organ via the renal vein and allowing it to escape through multiple minute perforations in the surface. (The minute holes are of no consequence to the subsequent function of the organ). Using phosphorous, MR it was found that when a standard preservation solution is used, kidneys stored for 24 h following 1 h of warm ischaemia show no evidence of high-energy metabolites, whereas kidneys treated in the same way but with the continuous administration of oxygen had measurable levels. This preservation technique has now been tested clinically and appears to offer the prospect of effective salvaging of organs that are capable of providing good renal function in the transplant recipient.

Another and related measurement that can be made by phosphorus MR is the determination of tissue pH by measuring the chemical shift of the inorganic phosphate peak. There is good evidence that a severe fall in pH, occasioned by anaerobic metabolism, is deleterious to subsequent function, and MR has been used to compare the effects of various buffers on the fall in pH during both warm ischaemia and cold storage. Part of the efficacy of the Ross and Marshall hypertonic citrate solution, which is commonly used for renal preservation, has been ascribed to its ability to maintain a higher pH during the preservation of human kidneys [9].

Clearly, many factors are involved in the effectiveness of hypothermic organ preservation, and energy status is only one of them. But phosphorous MR will have a continuing role in the development of new preservation solutions and possibly also in testing organs prior to transplantation.

Measurement of the Cryoprotectant Concentration

For a large number of cells it has been established that survival following freezing and thawing is crucially dependent upon the nature and concentration of the cryoprotective agent present. This factor interacts with other variables in the system, most importantly with cooling rate but also with warming rate. In the case of mouse haemopoietic cells, one of the first cell types to be rigorously studied, it was found that the optimum cooling rate moved to progressively lower values as the concentration of glycerol was increased, and that the peak survival characteristic of that optimum rate also increased. In the case of human red blood cells almost 100% survival can be obtained over a wide range of cooling rates, providing at least 3.3 *M* glycerol is present. The actual concentration of cryoprotective agent that can be used depends upon the extent to which the system in question tolerates replacement of water by the cryoprotectant. This in turn is determined partly by the osmotic effects of adding and removing the compound; consequently, optimisation of addition and removal procedures for the cryoprotectants in common use (glycerol and dimethyl sulphoxide are the most common) requires knowledge of the permeation kinetics and the temperature dependence of permeability of the cryoprotectants and of water. These values have been rigorously determined for rather few cell types, one such detailed study being that performed in our laboratory for human blood platelets. It was first determined that, in order to avoid osmotic damage to the platelets, their volume should not be reduced below 60% of initial volume nor increased above 130%. Measurement of the kinetics of glycerol penetration enabled a two-step addition procedure and a four-step removal procedure to be developed such that the cell volume remained within the permitted limits of +30% and −40% of isotonic volume [1]. These measurements were not actually made by MR methods, although they could have been, because other available methods were quite convenient. In the case of tissues, where other methods are very inconvenient, proton MR has been found to provide measurements of average cryoprotectant concentration non-destructively and quickly.

One tissue in which we are particularly interested is the cornea. This is the transparent window of the eye, and its function is crucially dependent upon integrity, not only of the special arrangement of the connective tissue which forms the transparent stroma, but also of the delicate layer of the cells that lines the internal surface. Corneas are transplanted in order to treat blindness due to opacity of the patient's own cornea. The results are extremely good so long as the transplanted cornea is viable, and rejection is much less of a problem in this situation than it is in many others because the cornea is avascular. Typical proton MR spectra for mixtures of the cryoprotectant dimethyl sulphoxide and water are shown in Fig. 2. Measurement of the peak areas produced a calibration curve that is identical to the predicted curve from the proton ratio for water and the cryoprotectant in the calibrating solutions. It then became possible to incubate corneas at the desired temperature for progressively increasing lengths of time in solutions of cryoprotectant, remove each cornea at the required time, rinse it in deuterium oxide, and then place it in the MR spectrometer to measure the ratio of cryoprotectant to water in the tissue. The attraction of this method is that it is entirely independent of osmotic phenomena since

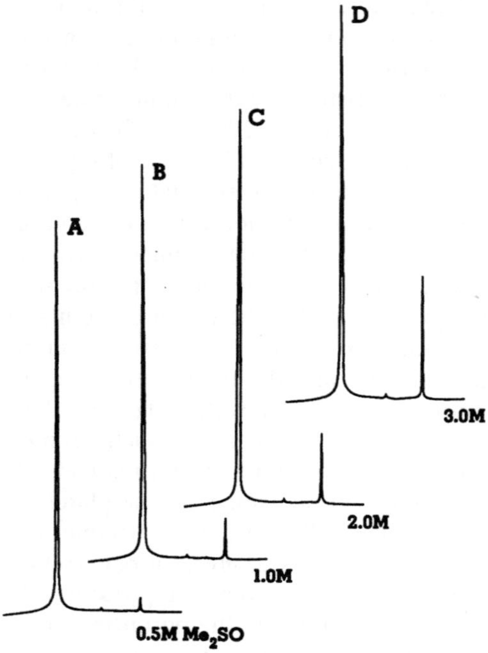

Fig. 2. Examples of MR spectra for Me_2SO solutions of different concentrations. The ratio of the areas under the peaks is in agreement with the theoretical proton ratios

it measures directly the variable that we need to know, namely, the concentration of cryoprotectant. In this particular case it was shown that at 0°C the concentrations equilibrate after 30 min of immersion, and that equilibration occurs more rapidly at 22°C and less rapidly at −10°C. On the basis of this knowledge it was possible to design a procedure for adding and removing the cryoprotectant which provided sufficient time for equilibration but did not expose the tissue unnecessarily to high concentrations at high temperatures. In other words, it was possible to incubate first at 0°C, then without unnecessarily prolonging exposure at this temperature to cool to −10°C and to continue permeation with a higher concentration of cryoprotectant at this lower temperature, which reduces chemical toxicity.

Analysis of these data has made it possible to determine effective permeability parameters for water and dimethyl sulphoxide in rabbit cornea at 22°, 0°, and −10°C [10]. Similar studies have been carried out for small arteries of the size that is required for coronary artery by-pass. (The model used was the rabbit common carotid artery). It was shown that equilibration is complete in 15–20 minutes at 2°C (see Fig. 3 [2]). Similar studies have now been carried out with other tissues and other cryoprotectants, for example, with kidney and smooth muscle tissue and with the cryoprotectant 2,3-butane-diol (unpublished studies). MR is proving to be an extremely convenient and powerful method for the determination of permeation kinetics for cryoprotectants in complex systems. It is of course limited

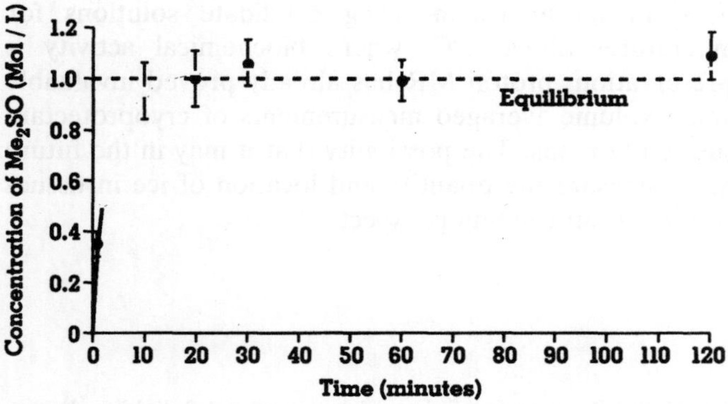

Fig. 3. Kinetics of Me_2SO permeation into the rabbit carotid artery at 2°C

to giving a volume-averaged measurement, but at the present state of knowledge of cryopreservation this is proving to be immensely valuable.

Location of Ice

We know from studies in a number of tissues, particularly smooth muscle and also renal tissue, that the location of any ice that is formed can be crucially important in determining subsequent injury to that tissue [7]. The methods used in these studies have been high-resolution techniques, but very slow and inconvenient: the most effective have been freeze-substitution and isothermal freeze-fixation followed by staining and either optical or electron microscopy. The possibility of using MR proton microscopy to obtain images of frozen tissues, even at the limited resolutions currently available, should be of considerable value in such studies. So far as this author is aware, this approach has not yet been explored.

Conclusions

MR spectroscopy has several attractions to the cryobiologist. First, it is non-invasive, and there are many circumstances in which the tissues that we deal with must be undamaged. Secondly, and connected to this, is the possibility of using spectroscopic examination as an individually predictive test for a specific organ, which offers the possibility of avoiding the transplantation of tissues and organs that subsequently prove to be ineffective. At the present time, this has not been realised in practice, no doubt due in part to the inadequate supply of organs for transplantation. The main value of MR spectroscopy has been in research comparing candidate solutions for preservation at temperatures above 0°C, where biochemical activity is continuing. In cryopreservation, proton MR has already proved invaluable as a means of obtaining volume-averaged measurements of cryoprotectant concentration in tissues and organs. The possibility that it may in the future be used, in addition, to measure the quantity and location of ice in tissues stored at low temperatures is an exciting prospect.

References

1. Arnaud FG, Pegg DE (1990) Permeation of glycerol and propan-1,2-diol into human platelets. Cryobiology 27:107–118
2. Bateson EAJ, Song YC (1991) Cryopreservation of arteries: I. Post-thaw smooth muscle function. Cryobiology 28:561–562
3. Chan L, French ME, Gadian DG, Morris PJ, Radda GK, Bore PJ, Ross BD, Styles P (1982) Study of human kidneys prior to transplantation by phosphorous nuclear magnetic resonance. In: Pegg DE, Jacobsen IA, Halasz NA (eds) Organ preservation. Basic and applied aspects. MTP Press, Lancaster, pp 113–119
4. English TA, Foreman J, Gadian DG, Pegg DE, Wheeldon D, Williams SR (1988) Three solutions for preservation of the rabbit heart at 0°C. J Thorac Cardiovasc Surg 96:54–61
5. Lovelock JE (1953) The mechanism of the protective action of glycerol against haemolysis by freezing and thawing. Biochim Biophys Acta 11:28–36
6. Mazur P (1963) Kinetics of water loss from cells at subzero temperatures and the likelihood of intracellular freezing. J Gen Physiol 47:347–369
7. Pegg DE (1987) Ice in tissues and organs. In: Pegg DE, Karow AM Jr (eds) The biophysics of organ cryopreservation. Plenum, New York, pp 117–140 (NATO ASI series A147)
8. Polge C, Smith AU, Parkes AS (1949) Revival of spermatozoa after vitrification and dehydration at low temperatures. Nature 164:666
9. Sehr P, Bore P, Thulborn K, Papatheofanis I, Chan L, Radda GK (1982) Tissue pH changes in renal preservation. In: Pegg DE, Jacobsen IA, Halasz NA (eds) Organ preservation. Basic and applied aspects. MTP Press, Lancaster, pp 109–112
10. Walcerz DB, Taylor MJ, Busza AL. Determination of the kinetics of permeation of dimethyl sulfoxide in isolated corneas and its temperature dependence from 22°C to −10°C using nuclear magnetic resonance spectroscopy. Exp Eye Res (submitted)

Image-Guided Volume-Selective Magnetic Resonance Spectroscopy and Relaxometry: Application to the Porcine Kidney Perfused with Perfluorocarbons*

G. Schnur[1]

Introduction

Image contrast in magnetic resonance (MR) tomography is governed largely by the relaxation times T1 and T2 of the tissue. In imaging procedures, relaxation enters in an indirect way. On the basis of volume-selective signals one has the possibility to measure precisely parameters which are suitable for the characterization of tissue. We have therefore developed pulse sequences for the localized and spectroscopically resolved determination of precise relaxation times.

First a method for volume-selective and spectroscopically resolved relaxometry in MR tomography and second a method for retuning a ^{1}H bird-cage resonator from ^{1}H to ^{19}F resonance is presented. As an example the applications to ^{19}F imaging and volume-selective ^{19}F spectroscopy and relaxometry of perfluorocarbon emulsions in kidney tissue are reported.

Pulse Sequences for Localized Spectroscopy and Relaxometry

The pulse sequences serve the volume-selective measurement of the MR relaxation times T1 and T2 of resolved (or edited) spectral lines. Figure 1 shows the radiofrequency (RF) and field-gradient pulse schemes. The pulse sequences are subdivided into a relaxation part and a subsequent localization procedure. For localization the pulse sequences produce volume-selective echoes by the aid of three slice-selective pulses (VOSY sequence) [1]. For T1 measurements, the inversion recovery or saturation recovery methods are used with the 90° detection pulse replaced by the VOSY sequence. The transverse relaxation time T2 is determinded on the basis either of the Hahn 90°/180° pulse sequence or the Carr-Purcell-Gill-Meiboom (CPGM) sequence. The (last) 180° pulse in these case also serves the excitation of the

* All work was carried out at the Sektion Kernresonanzspektroskopie, Universität Ulm, D-89081 Ulm, FRG.
[1] Siemens AG, UB Medizinische Technik, D-91052 Erlangen, FRG

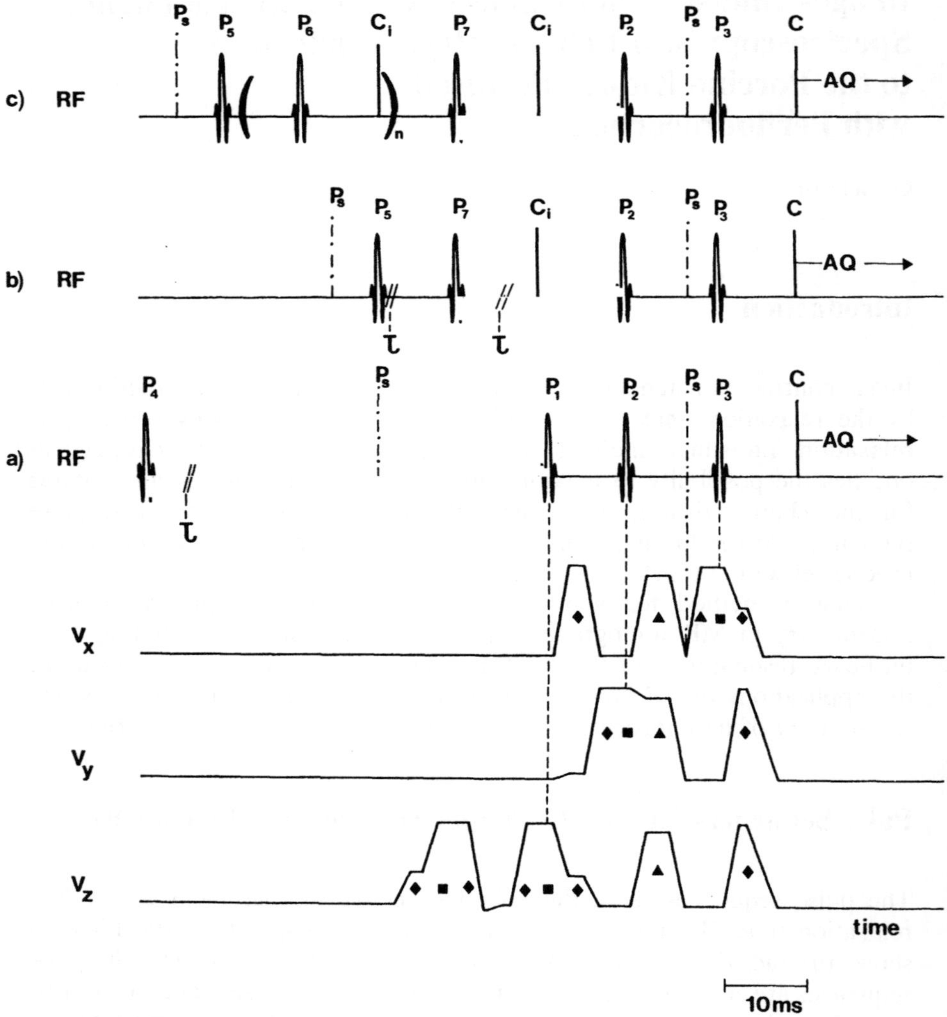

Fig. 1. RF and field-gradient pulse sequences for localized relaxometry. *a*, T1; *b*, T2 (Hahn); *c*, T2 (CPGM). P_1, P_2, P_3, VOSY pulses; P_4, P_5, pulses preparing the initial nonequilibrium state; P_6, P_7, 180° refocusing pulses; P_s, selective saturation or suppression pulses. C_i, intermediate echo; *C*, volume-selective transfer echo; τ, variable relaxation delay; *AQ*, acquisition interval. The control voltages (V_x, V_y, V_z) of the field gradients are drawn in scale as applied to the gradient power supply. ▲, Spoiling of undesired coherences; ◆, balance of dephasing/rephasing intervals; ■, selection of slices

first slice so that the (next) echo is already slice selective. It therefore replaces the first VOSY 90° excitation pulse and only two additional VOSY pulses are needed for the production of a volume-selective signal C. The relaxation process is of course continued during the intervals of the localization pulses. Thus the acquired signals are reduced by a constant factor independent of the length of the proper relaxation delay.

Test of Pulse Sequences for Localized Spectroscopy and Relaxometry

The experimets were carried out with a Bruker Biospec 47/40 tomograph operating at 200 MHz proton resonance [3]. The room temperature bore diameter of the magnet was 40 cm, but a gradient tube with 30 cm inner diameter was used to reduce the induction of eddy currents. The ramp time of the gradient pulses was 2 ms. The typical gradient value for selection of voxel volume of 1 ml was 2.5 kHz/cm (in proton Lamor frequency units). The slice-selective pulses were filtered sinc pulses with a length of 2 ms and a bandwidth of 2.5 kHz.

The reliability of the pulse sequence for the localized determination of T1 and T2 was tested by the aid of aqueous $CuCl_2$ solutions in the concentration range of 4.0×10^{17}–1.6×10^{19} ions/ml. T1 and T2 were measured selectively in a volume element of 1 ml as well as with the conventional techniques without localization. Volume-selective and unlocalized data coincide within the experimental error of a few percent (Fig. 2).

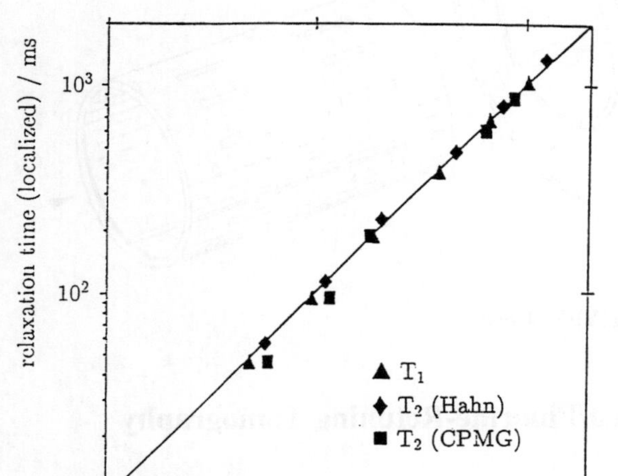

Fig. 2. Relaxation times T1 and T2 of aqueous $Cucl_2$ solutions measured localized in a volume element of 1 ml versus relaxation times measured with conventional techniques without localization. ▲, spin-lattice relaxation time T1; ◆, transverse relaxation time T2 (measured with Hahn method); ■, transverse relaxation time T2 (measured with CPGM method)

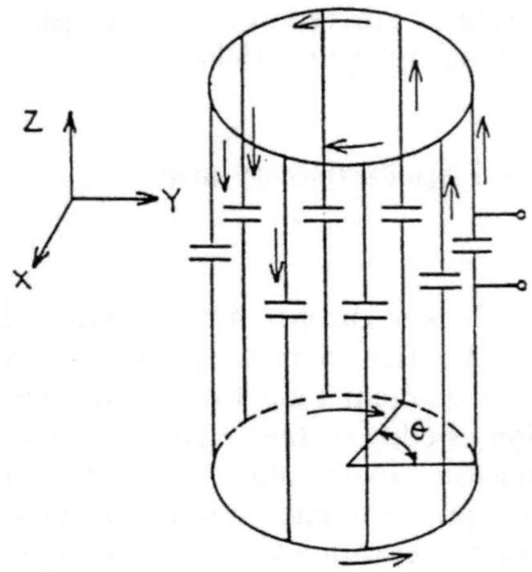

Fig. 3. The low-pass version of a bird-cage resonator consists of circular end rings and straight segments with capacitors. (Static magnetic field in *z* direction. Rotating rf magnetic field in *x*,*y* planes)

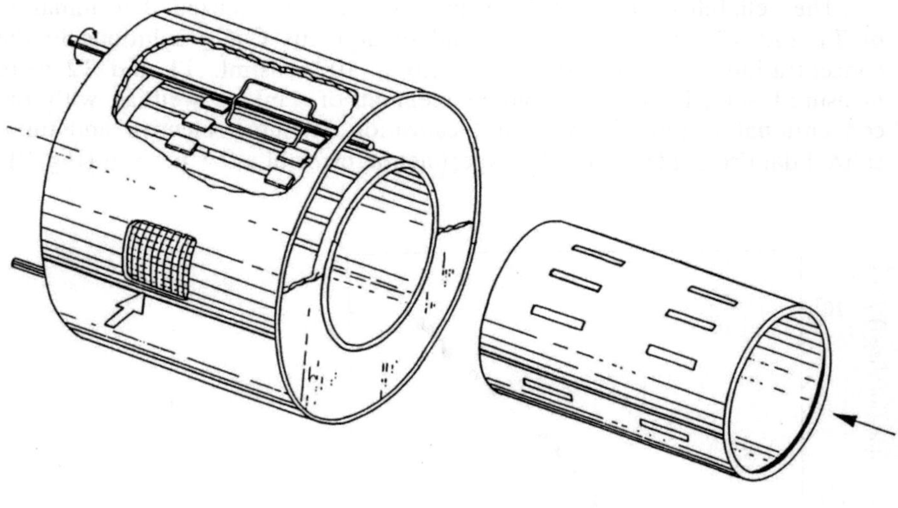

Fig. 4. Bird-cage resonator and HYFY sleeve

Principle of the Hydrogen/Fluorine-Retuning Tomography

An MR bird-cage resonator [2] as a constituent of the Bruker Biospec BMT 47/40 tomography system consists of a group of $\lambda/2$ rod aerials in a "bird-cage" arrangement (Fig. 3). The construction design requires fixed values of capacitance. The fine-tuning is performed by the aid of a resonant coupling circuit. The tuning range is therefore limited to a few percent. In other words, the proton resonator cannot be tuned to fluorine resonance (or vice versa).

For retuning the proton resonator to both resonances we used the following copper strip tube method. The capacitance of the rod antennas is increased by copper strips mounted on a plastic tube (Fig. 4). After inserting the retuning tube in the resonator, the strips are located adjacent to the shortening capacitors of the rod aerials. They therefore act as additional capacitors. For this purpose the rod antennas are suitably formed as strips as is in the case with the Bruker resonator. The fine-turning can be carried out by the usual "tuning" and "matching" elements of the probe head.

Test of the Hydrogen/Fluorine-Retuning Technique

The method was tested with the Bruker 200 MHz proton resonator [3]. The inner diameter was 150 mm. Twenty-eight copper strips of 12 × 44 mm were mounted on a polymethylmetacrylate retuning tube. With the retuning tube a quick shift in the resonance frequency of the resonator to the ^{19}F resonance frequency of 188 MHz is possible. No change of probe head or probe is necessary. The bandwidth, Q factor, and RF field homogeneity of the resonator with and without retuning tube were measured. In the fluorine-tuned state, the bandwidth of the empty resonator was somewhat

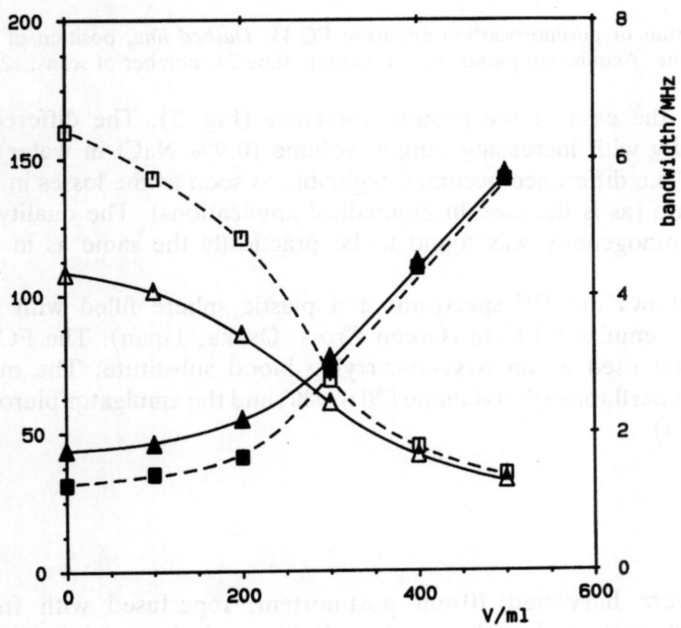

Fig. 5. Quality (□, △) and bandwidth (■, ▲) at proton (*dashed line*) and fluorine (*solid line*) resonance versus sample volume (physiologic salt solution)

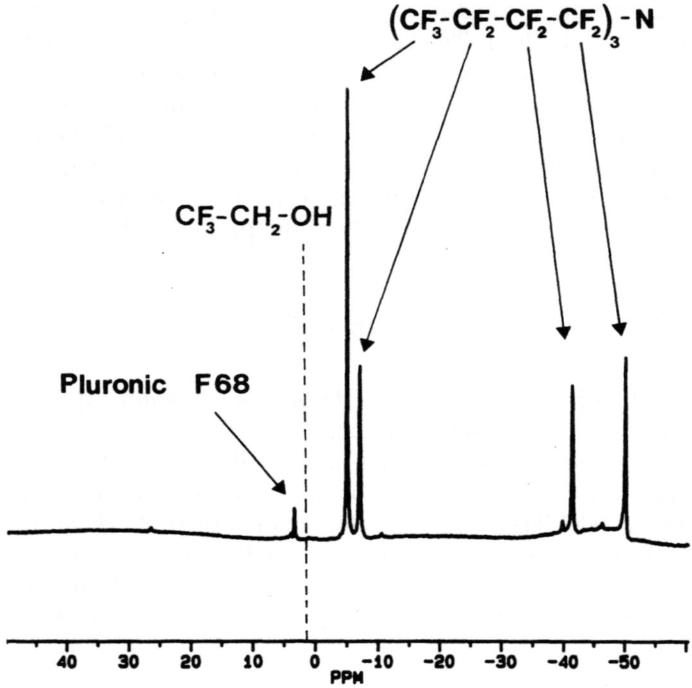

Fig. 6. ^{19}F spectrum of perfluorocarbon emulsion FC-43. *Dashed line*, position of the trifluoroethanol line. Acquisition parameters: repetition time 2 s, number of scans, 12

higher than in the case of the proton resonance (Fig. 5). The difference decreases rapidly with increasing sample volume (0.9% NaCl in water) in the resonator. The difference becomes negligible as soon as the losses in the sample dominate (as is the case in biomedical applications). The quality of the RF field homogeneity was found to be practically the same as in the proton case.

Figure 6 shows the ^{19}F spectrum of a plastic sphere filled with the perfluorocarbon emulsion FC-43 (Green Cross, Osaka, Japan). The FC-43 emulsion can be used as an oxygen-carrying blood substitute. The main constituents are perfluorotributylamine (20 w/v%) and the emulgator pluronic F-68 (2.56 w/v%).

Experiments

Pig kidneys were harvested 10 min postmortem, reperfused with fresh pig blood, and preserved with 4°C EuroCollins solution. After 4 h of cold storage the kidneys were reperfused with oxygenated perfluorocarbon emulsion (FC-43).

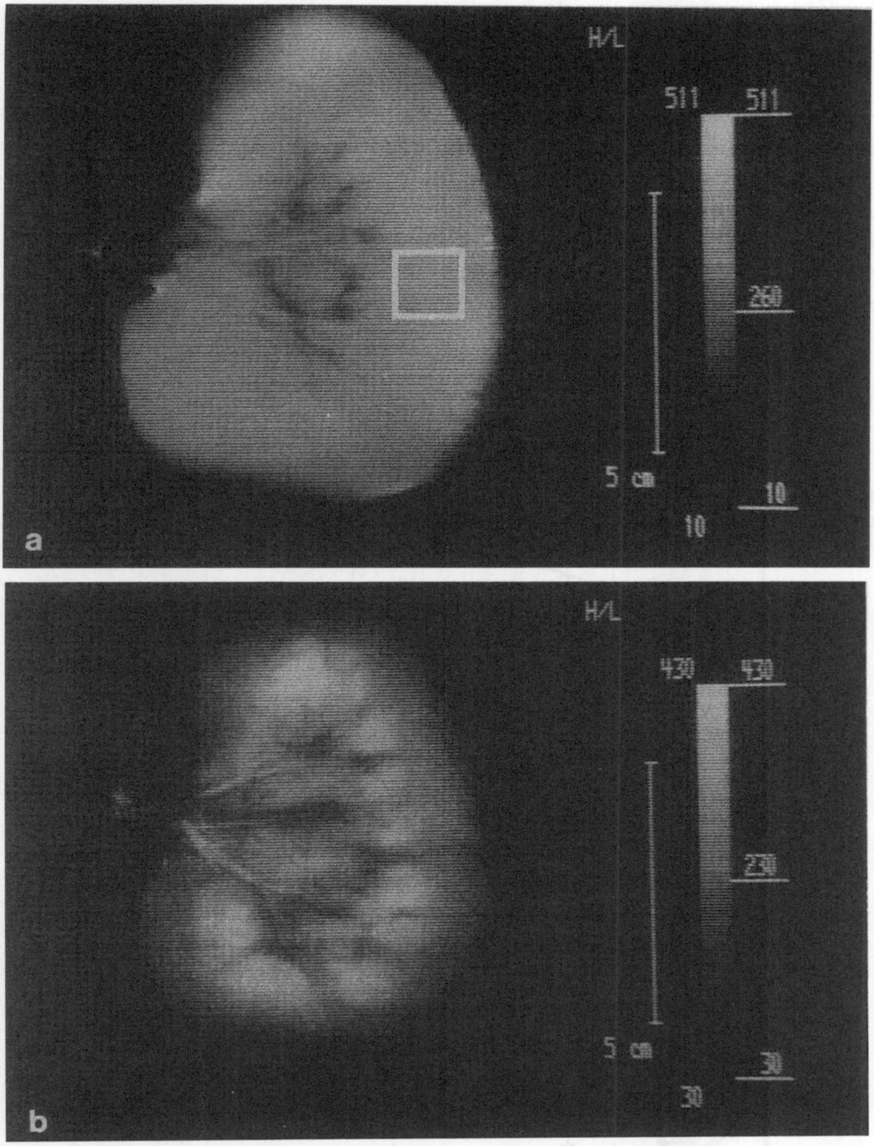

Fig. 7. ^{1}H (**a**) and ^{19}F (**b**) images of the same slice of porcine kidney perfused with perfluorocarbon emulsion. ^{19}F acquisition parameters: TR = 1.2 s, TE = 34 ms, number of scans 8, number of pixels 256 × 256, total imaging time 41 min. The fluorine image is based on the lines between 0 and −10 ppm (Fig. 6)

The regeneration of ATP and PCr was monitored with ^{31}P spectroscopy at 81 MHz using a double tuned (81/200 MHz) surface coil.

The perfusion was controlled with the hydrogen/fluorine-retuning technique [3]. Figure 7 shows the proton and fluorine images. The volume of the blood vessels was about 7% of the organ volume. In the completely FC-43 perfused state the content of perfluorotributylamine in tissue was only 1.4%. Therefore the thickness of the slices was chosen to be 10 mm, and eight scans were accumulated for the fluorine images. Note that the images were recorded without removing the organs, so that identical slices are reproduced.

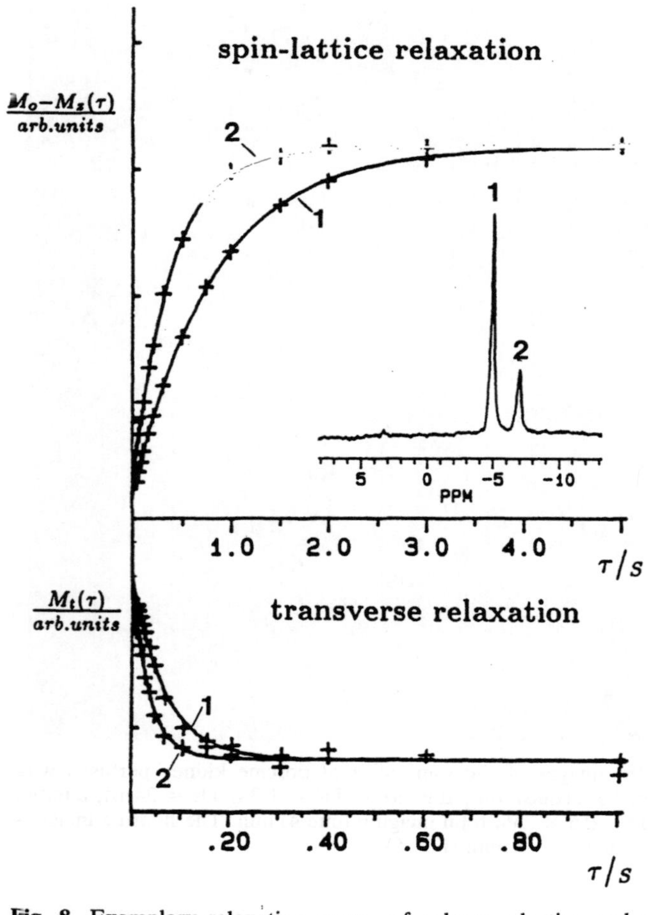

Fig. 8. Exemplary relaxation curves of volume-selective and spectroscopically resolved ^{19}F relaxometry in porcine kidney perfused with perfluorocarbon emulsion FC-43. The spectral lines evaluated are indicated. Acquisition parameters: excited volume $(1.5\,cm)^3$, repetition time 5 s; number of scans 8

Table 1. Localized relaxation times of perfluorocarbon emulsion FC-43 in porcine kidney

	T1 (s)	T2 (s)
CF_3 line	0.86 ± 0.03	0.068 ± 0.005
CF_2 line	0.41 ± 0.02	0.040 ± 0.003

Acquisition parameters: excited volume $(1.5\,cm)^3$, repetition time 5 s, number of scans 8. The T2 values are not corrected for the influence of indirect spin-spin coupling.

It is possible to localize the volume of interest on the basis of the proton image. The volume-selective recorded fluorine spectrum of the indicated voxel is shown in Fig. 8. The large chemical shifts of the spectral lines lead to a correspondingly large range in spatial shifts of the exited volume elements. Therefore we restricted ourselves to a section of the fluorine spectrum.

The local fluorine relaxation times T1 and T2 in the voxel indicated in Fig. 7a was measured by the aid of the pulse sequences shown in Fig. 1. The quality of the relaxation curves is demonstrated in Fig. 8; Table 1 summarizes the fitted relaxation times.

Since different saturations of perfluorocarbon with O_2 causes different T1 relaxation times of fluorine, it should be possible to determine (indirectly) the O_2 concentration in tissue perfused with perfluorocarbons.

Conclusion

The experiments carried out in this study demonstrate the capability of the proposed localized measuring techniques. The hydrogen/fluorine-retuning technique permits the recording of fluorine images, spectra, volume-selective spectra, and relaxation times with a proton resonator. Thus the localization of the fluorine investigation can be performed on the basis of the proton images.

Acknowledgements. I thank R. Kimmich, R. Lietzenmayer, and all my colleagues at the University of Ulm for cooperation and discussions in the course of the work.

References

1. Kimmich R, Höpfel D (1987) Volume-selective multipulse spin-echo spectroscopy. J Magn Reson 72:379
2. Hayes C, Edelstein W, Schenk JF, Mueller OE, Eash M (1985) An efficient, highly homogeneous radiofrequency coil for whole-body NMR imaging at 1.5 T. J Magn Reson 63:622
3. Schnur G, Kimmich R, Lietzenmaer R (1990) Hydrogen/fluorine retuning tomography. Applications of ^{1}H image-guided volume-selective ^{19}F spectroscopy and relaxometry of perfluorocarbon emulsions in tissue. Magn Reson Med 13:478

Diffusion Magnetic Resonance Microscopy of Kidney Preservation*

M. Neeman[1], L.J. Gibel[2], S.M. Smith[2], A. Hartford[2], W. Sterling[2], and L.O. Sillerud[1]

Introduction

Cellular swelling, due principally to the influx of sodium and water, has long been recognized as a cause of tissue injury during prolonged storage of transplantable kidneys [1]. Reliable storage for up to 48 h was made possible through the use of hyperosmotic solutions with intracellular electrolyte composition and glucose (EuroCollins solution). Improved function in the early postoperative period appears to have been achieved by incorporating osmotic agents that are more effective cellular impermeants (Belzer's solution).

^{31}P magnetic resonance (MR) spectroscopy has been used to quantify the pH and the depletion of high-energy phosphate metabolites during renal preservation [2, 9]. Comparison with early posttransplantation function demonstrates conflicting findings. These discrepancies may be explained by noting that cellular injury during hypothermic preservation may be relatively independent of energy stores. MR imaging studies on post-transplant kidney function appear to show a correlation between kidney function and cortex/medullary differentiation [4].

MR microscopic images of excised rat kidneys show remarkable resolution. In studies on the dependence of image contrast on T1 relaxation [3], T1 was found to increase from the cortex to the inner medulla. One of the goals of this work was to determine the potential contribution of diffusion contrast to kidney images. Diffusion MR imaging is a very sensitive marker of cellular packing and ordering [6] and has shown superior contrast in early detection of edema in the brain following ischemic injury [5]. In this work the sensitivity of MR diffusion mapping to intra- and extracellular edema was explored in rat kidneys during hypothermic preservation.

---

* This work was conducted under the auspices of the United States Department of Energy.

[1] Life Sciences Division, Los Alamos National Laboratory, Los Alamos, NM 87545, USA
[2] Department of Urology, The University of New Mexico, Albuquerque, NM 87131, USA

Materials and Methods

Kidney Retrieval. Mature male Sprague Dawley rats (300–500 g) were anesthetized with sodium pentabarbitol (16.5 mg intraperitoneal). Rat kidneys were removed after a flush in situ using chilled preservation solution, in a manner analogous to the standardized technique for human cadaver renal retrieval [10]. Ice-chilled preservation solution was delivered through the aortic cannula using a peristaltic pump at a rate of 10 ml/min for 10 min. The venous circulation was vented by incising the left renal vein adjacent to the vena cava. During perfusion the abdomen was filled with crushed ice. Three types of flushing solutions were tested (three animals for each solution): lactated Ringer's solution (Kendal McGaw, USA), EuroCollins solution (Fresenius, FRG) and Belzer's VIASPAN preservation solution (DuPont Pharmaceuticals, USA). One kidney from each animal was placed into a 20-mm MR tube for MR studies. The second kidney was used for histological studies, where slices 1 mm thick were taken from the kidney and were immediately immersed in 10% formalin at 30 min, 4 h, 8 h, and 24 h after flush.

MR Imaging. MR diffusion experiments were performed at 4°C on a Bruker wide-bore AM 400 spectrometer equipped with a microimaging accessory, as described previously [7, 8]. Consecutive sets of four 256 × 256 spin-echo images were obtained with incrementation of the diffusion gradient (2.4 h acquisition per set of four images). The corresponding b values [5, 6], including the contribution due to the imaging gradients [7, 8], were 12380, 24200, 41780, and 65080 s/cm^2 with a diffusion time of 24 ms. T_2-weighted images were obtained with TE of 60 ms and TR of 2 s. A slice thickness of 0.5 mm was obtained using a slice-selective 90° sinc pulse (10 ms). The in-plane resolution was 43 × 43 μm.

Results and Discussion

MR microimaging provides a very detailed view of the structure and morphology of kidneys. The images clearly show the cortical and medullary regions, collecting ducts, arteries, and kidney sinus (Fig. 1). A set of images with incremented diffusion gradients was obtained for each time point to separate the effect of T_2 and diffusion and to detect changes in water diffusion.

Kidneys stored with lactated Ringer's showed progressive prominence of fluid-filled structures at the cortical-medullary junction. The area of a cross-section through these structures increased by approximately a factor of 2 during 10 h of storage in lactated Ringer's solution. An accurate measurement of the rate and kinetics of this dilation were beyond the resolution

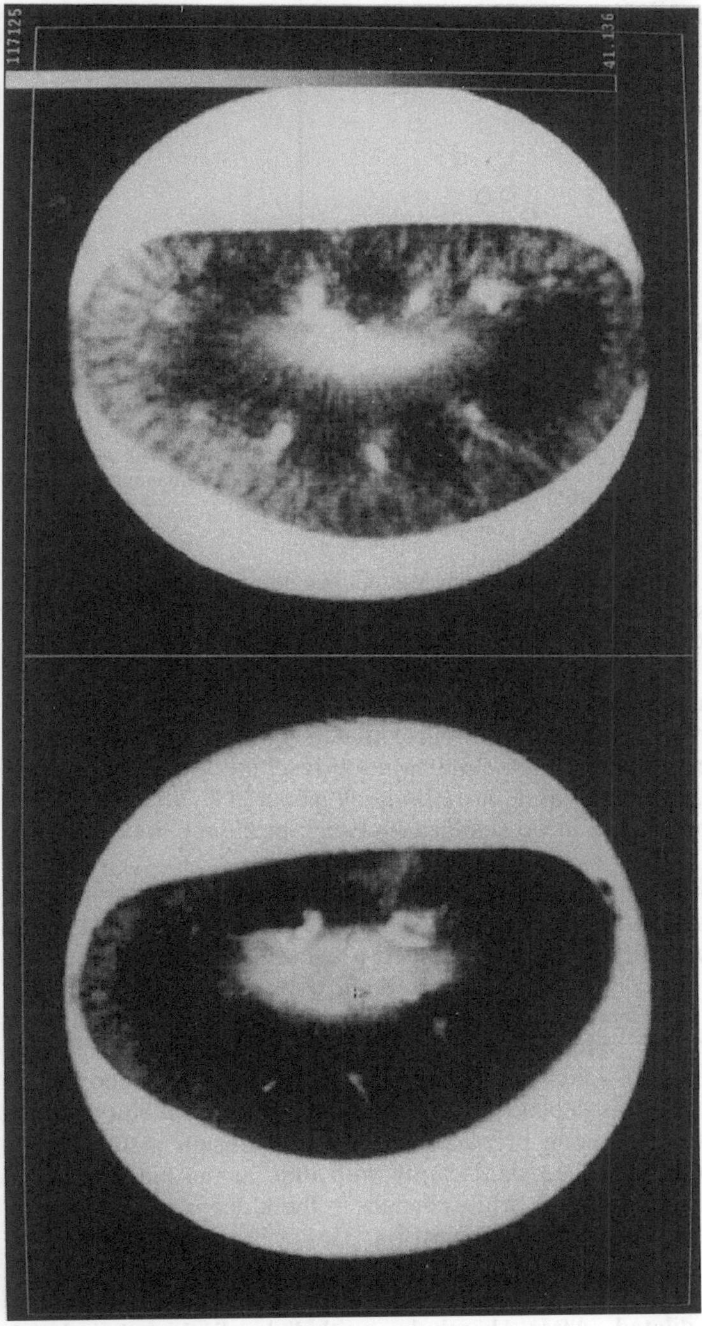

Fig. 1. MR images of a rat kidney flushed with lactated Ringer's solution. *Left*, 2.4h after kidney retrieval; *right*, 24h after kidney retrieval

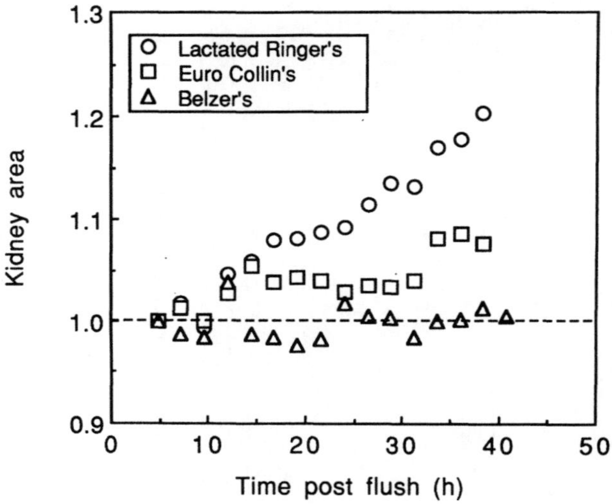

Fig. 2. Water uptake during hypothermic preservation. The area of the kidney MR image was measured for the same slice at different times following kidney retrieval and scaled with respect to the first image at 5 h after flush

of the images. Correlation with histological specimens to determine the nature of these structures is pending.

The effect of the preservation solution on kidney swelling was very dramatic (Figs. 1, 2). Perfusion and storage of the kidneys with lactated Ringer's solution resulted in rapid swelling of the kidney. The area of the kidney in the image increased by approximately 20% after 24 h. Storage of kidneys preserved with EuroCollins solution resulted in a slower rate of kidney swelling, with the area increasing by about 5% after 24 h. The Belzer's solution resulted in no significant change in kidney area following 40 h of storage.

The effect of the preservation solution on signal intensity was observed clearly in the cortical and medullary regions of the kidneys (Fig. 1). The ratio of signal intensity in the cortex to the intensity in the kidney sinus showed a clear time dependence for the three preservation solutions. The ratio increased significantly in the kidneys stored with lactated Ringer's solution and did not change with time for the two other solutions. The difference in the initial value for the cortex-to-sinus ratio depended mainly on T_2 relaxation and was dependent on the amount of residual blood, which resulted in variability among preparations. Water diffusion in the kidney cortex and medulla increased significantly with time in the kidneys stored with lactated Ringer's solution, while changes in the kidneys stored with the two other solutions were not significant (Fig. 3).

Histological specimens of kidneys perfused with lactated Ringer's solution showed severe damage as early as 15 min after perfusion. Proximal tubules became dilated. After 4 h tubular epithelial cells lost their brush border, and there was swelling and vacuolization of the cytoplasm. All these changes increased in severity with time, and after 20 h following perfusion

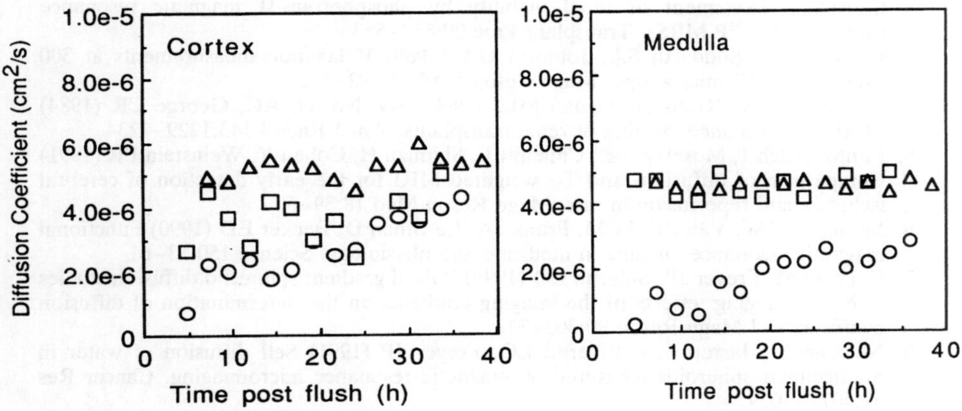

Fig. 3. Water diffusion in preserved kidneys. The diffusion coefficients for water in the cortical and medullary regions of the kidney are presented for representative preparations of kidneys flushed and stored with the three preservation solutions (*circles*, lactated Ringer; *squares*, Euro Collins'; *triangles*, Belzer)

there was extensive disruption of the tubular epithelium with sloughing of the apical portion. In contrast, histological specimens of kidneys perfused with EuroCollins solution or Belzer's solution evidenced much milder changes. Tubular dilation was focal and did not increase with perfusion time. Tubular vacuolization and brush border loss were focal and mild, even after 24 h.

In conclusion, the degradation of rat kidneys undergoing hypothermic preservation with lactated Ringer's solution correlated well with clearly observable findings by MR diffusion imaging. Findings included reduced diffusion coefficients in both the cortical and medullary areas when compared to EuroCollins and Belzer stored kidneys at the early time points. In addition, lactated Ringer kidneys demonstrated a progressive rise in diffusion coefficient and in cross-sectional area size. While these findings represent demonstrable alterations in an unfavorable mode of kidney preservation, they suggest the potential of MR imaging in noninvasive assessment of cellular swelling and extracellular space edema. The utility of diffusion imaging in combination with pH and high-energy phosphate determination by ^{31}P MR spectroscopy deserves further study.

References

1. Belzer FO, Hoffman RM, Southard JH (1978) in "Kidney Preservation" Symp. Organ Transplant. Surg Clin North Am 58:261
2. Bretan PN, Baldwin N, Novick AC, Ng TC, Majors A, Stowe N, Streem S, Steinmuller D, Go R, Meany T (1988) Preliminary clinical experience with pre-

transplant assessment of renal viability by phosphorous-31 magnetic resonance spectroscopy (^{31}P MRS). Transplant Proc 20:852–853
3. Dockery SE, Suddarth SA, Johnson GA (1989) Relaxation measurements at 300 MHz using MR microscopy. Magn Reson Med 11:182–192
4. Geisinger MA, Risius B, Jordan ML, Zelch MG, Novick AC, George CR (1984) Magnetic resonance imaging of renal transplants. Am J Radiol 143:1229–1234
5. Mintorovitch J, Moseley ME, Chileuitt L, Shimizu H, Cohen Y, Weinstein PR (1991) Comparison of diffusion- and T_2- weighted MRI for the early detection of cerebral ischemia and reperfusion in rats. Magn Reson Med 18:39–50
6. Moonen CTW, Van Zijl PCM, Frank JA, Le Bihan D, Becker ED (1990) Functional magnetic resonance imaging in medicine and physiology. Science 150:53–61
7. Neeman M, Freyer JP, Sillerud LO (1990) Pulsed gradient spin echo diffusion studies in NMR imaging: effects of the imaging gradients on the determination of diffusion coefficients. J Magn Reson 90:303–312
8. Neeman M, Jarrett KA, Sillerud LO, Freyer JP (1991) Self diffusion of water in multicellular spheroids measured by magnetic resonance microimaging. Cancer Res 51:4072–4079
9. Pomer S, Hull WE, Röhl L (1988) Assessment of renal viability for transplantation by high field ^{31}P NMR. Transplant Proc 20:899–901
10. Simmons RL, Finch ME, Ascher NL, Najarian JS (1984) Multiple organ donation from a cadaver, In: Simmons RL et al. (eds) Manual of vascular access, organ donation and transplantation. Springer, Berlin Heidelberg New York, pp 103–143

Magnetic Resonance Microscopy of the Mouse Kidney

H.-M. Cheng[1], L. Garrido[2], E. Brown[1], and J.B. Aguayo[1]

Introduction

Clinical magnetic resonance (MR) imaging is now an established routine procedure. With conventional spin-echo, the in-plane image resolution can be as high as 313 × 313 μm with a slice thickness of 3 mm [5]. Beyond this, special approaches must be used, for example, reducing the field-of-view and/or increasing field gradient strengths, both of which can be achieved with the implementation of special software and hardware. The signal-to-noise ratio can be further improved by imaging at high magnetic fields in conjunction with small radiofrequency coils. High-contrast microimages can therefore be acquired. Using a combination of these techniques, Aguayo and coworkers [1] have demonstrated MR microscopy with a modified high-field MR spectrometer. They were able to obtain images of a frog egg with an in-plane resolution of 10 × 13 μm. This method has been extended to the examination of enucleated human and animal eyes and tumors in mice [2, 3, 10] and mapping of water and ^{13}C metabolites in the diabetic lens [4]. In vivo microscopy of acute tubular necrosis in the rat kidney resulting from toxicity has been carried out by Farmer et al. [7]. They implanted inductively coupled coils around the kidneys, and with which voxels of 117 × 117 × 2000 μm were obtained with spin-echo that allowed visualization of the cortex, the inner and outer medulla, and the major vasculature. In this article, we demonstrate microimaging of the mouse kidney at 35 × 35 × 500 μm resolution and investigate the hydrodynamic behavior of water in the kidney at the same time.

[1] Howe Laboratory of Ophthalmology, Harvard Medical School, and Radiology Service, the Massachusetts Eye and Ear Infirmary, 243 Charles Street, Boston, MA 02114, USA
[2] Department of Radiology, Harvard Medical School, and the Massachusetts General Hospital Imaging Center, Charlestown Navy Yard, Charlestown, MA 02129, USA

Materials and Methods

CD-1 albino and C-57 black mice (around 50 g body weight) were obtained from Charles River Breeding Laboratories (Wilmington, MA). Either the whole body was positioned in the imaging probe, or the excised kidneys were placed in 10-mm MR tubes before loading into the probe. Excised kidneys measured around 5 × 7 × 12 mm.

The MR microscope is a modified Bruker MSL400 MR spectrometer operating at 400 MHz for proton. It has a bore size of 8.9 cm and is equipped with microimaging attachments including power supplies for the field gradients, an imaging probe with 5-cm diameter gradient coils, resonators of different sizes, a pulse programmer, and image processor and display. Slotted tube resonators of 25- and 10-mm diameter were installed in the probe for imaging of the whole body and excised kidneys, respectively.

The imaging sequences included conventional spin-echo and pulse gradient spin-echo (PGSE) intended for the analysis of morphology and water motion, respectively. The PGSE imaging sequence [6] was a modification of that of Stejskal and Tanner [8] with two equal gradient pulses starting immediately after slice selection and which are separated by the 180° refocusing pulse. Gradient recalled echo sequences also were tested; however, the susceptibility effect from unidentified contaminants in the kidneys was too great to permit reasonable image acquisition. MR parameters and image resolution are indicated in figure legends.

Results and Discussion

The purposes of this study were essentially: (a) to show kidney images in micron resolution, (b) to test water diffusion imaging sequence, and (c) to generate background information for future clinical application.

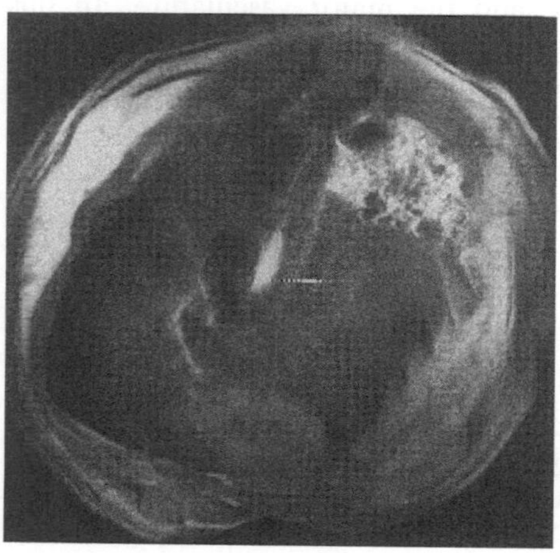

Fig. 1. An axial T1-weighted image of the abdomen. MR parameters included TR/TE = 500/5.3 ms and 256 × 256 display with 16 signal averages. The in-plane resolution was 110 × 110 μm and slice thickness 1 mm. The two kidneys can be seen at 5 and 7 o'clock, respectively

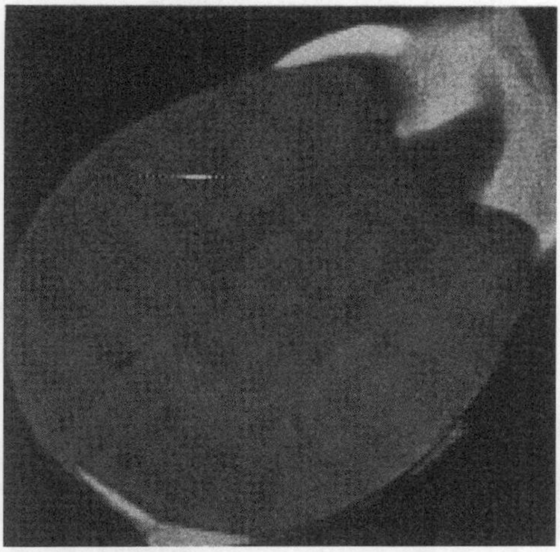

Fig. 2. T1-weighted axial image of an excised kidney. MR parameters were identical to those in Fig. 1, except that the voxel size was 35 × 35 × 500 μm

Figure 1 shows an axial section through the midline of the mouse kidneys in situ. These kidneys were 4–5 mm in diameter in midsection. The kidney on the right overlapped with the abdominal skin and therefore was partially obscured. The kidney on the left is clearly seen, and its cortex and medulla are discernible. Image resolution at this level (110 × 110 × 500 μm), although similar to the previous rat study [7], does not permit visualization of fine structures in the mouse kidney; for this, higher resolution must be attained:

Figure 2 is the axial image of an excised mouse kidney. The field-of-view was reduced to 9 mm, which allowed an in-plane resolution of 35 × 35 μm (i.e., with 256 each frequency data points and phase-encoding steps). The bright regions at the top and bottom of the kidney are fatty tissues. The kidney itself can be clearly divided into the medulla, which appears dark, and the cortex, which is somewhat inhomogeneous with several black spots arranged radially that correspond to collecting ducts.

Diffusion images are shown in Fig. 3. Water diffusion in suspended renal proximal tubules has previously been examined with spectroscopy [9]; however, diffusion imaging of the kidney has never been demonstrated effectively. We used the PGSE sequence designed to record the spatial information of water self-diffusion [6, 8]. The images are best appreciated in connection with the conventional spin-echo image (Fig. 2). With zero gradient (Fig. 3a), the medulla and the medulla-cortex interface appear brighter than the cortex; because of the longer TE than that used for Fig. 2 (with the same TR) and the inherently short T2 values of the kidneys, the increasing brightness was apparently a T1 effect. When the x-gradient is turned on, the part of the medulla that appears dark indicates the orientation of aqueous channel/pathway in the x-direction (Fig. 3b). When the z-gradient is on, a new area of the cortex now appears dark, suggesting channel orientation in this direction (Fig. 3c); however, this new area is wedged-

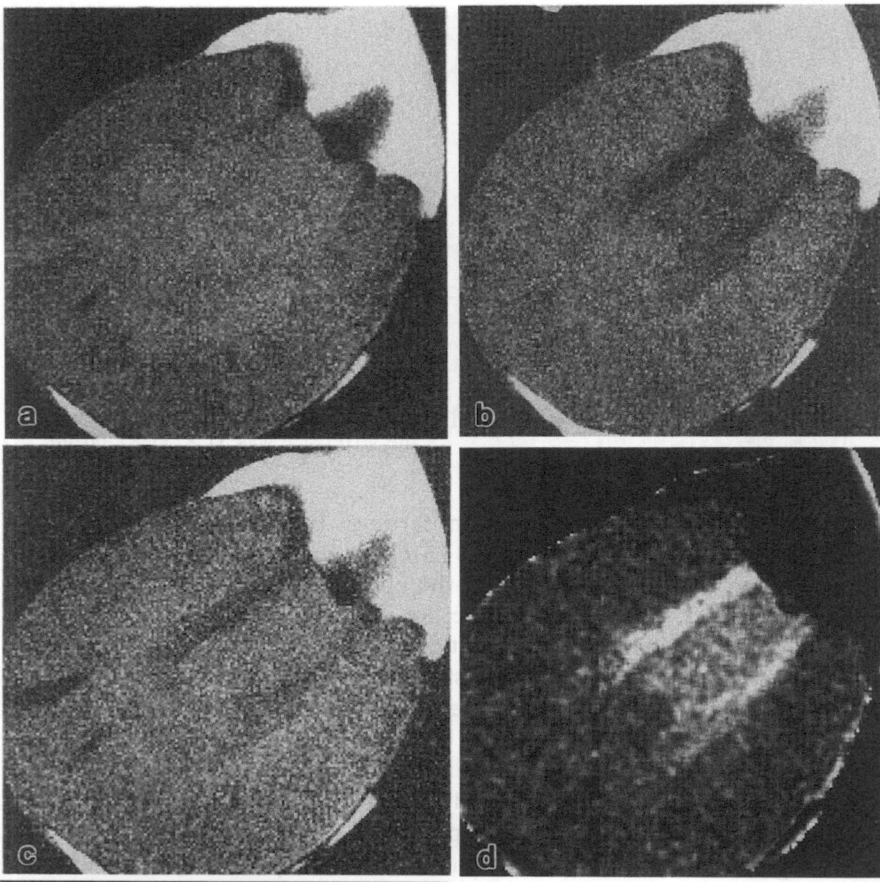

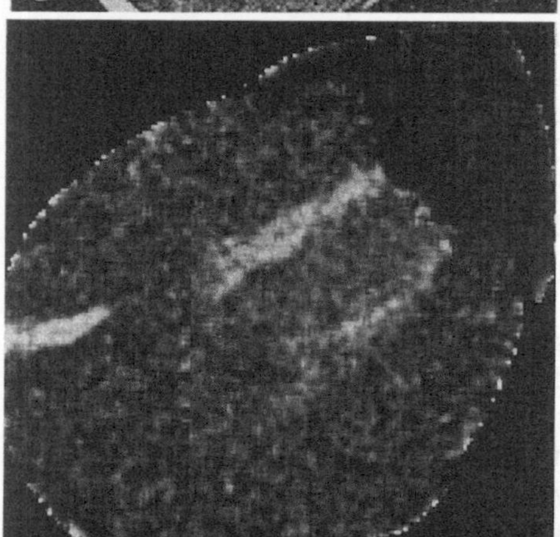

Fig. 3. Diffusion imaging of the same kidney as in Fig. 2. MR parameters included TR/TE = 500/17 ms with eight averages; the duration of each of the two pulsed gradients was 5 ms, and the time between the beginning edges of the two pulses was 7 ms. The voxel size was 35 × 35 × 500 μm. The background image (**a**) was acquired with zero gradients. The field gradient strength was 20 G/cm along either (**b**) the x-gradient (left → right) or (**c**) the z-gradient (along the long axis of the kidney, perpendicular to the image plane). Diffusion maps with x- and z-gradients are recorded in (**d**) and (**e**), respectively; they were calculated from images acquired with 0 and 20 G/cm gradients

shaped, which may be from infarct or trauma. On the other hand, most of the cortical water channels appear randomly oriented. Diffusion maps (Fig. 3d, e) clearly show high- (bright) and low-diffusion regions. The high-diffusion regions, especially the medulla, are with diffusion coefficients of $2-3 \times 10^{-5}\,\mathrm{cm}^{-2}\,\mathrm{s}^{-1}$, identical to that of water, indicating no impedance in these regions. This result is in agreement with the water-concentrating function of the medulla. On the other hand, the cortex has a coefficient of $0.5 \times 10^{-5}\,\mathrm{cm}^{-2}\,\mathrm{sec}^{-1}$.

These examples illustrate microimaging that provides detailed morphological information as well as the motion of water. Because of the specific function of the kidney, information on water diffusion is particularly important.

In summary, MR microscopy is useful at present for small animal studies on models such as tumors, diabetes, and transplants. Its clinical application to humans requires high-field (e.g., 1.5–4 T) body scanners with small gradient coils and surface radiofrequency coils. Diffusion imaging further requires the total control of motion, which can be achieved by employing fast imaging. When implemented, MR microscopy will prove invaluable in clinical diagnosis, for it represents a truly noninvasive biopsy procedure.

Acknowledgements. This project was supported in part by EY07620 (to H.M.C.) awarded by the National Eye Institute, National Institutes of Health, Bethesda, MD, and in part by a feasibility grant from the American Diabetes Association (to H.M.C.).

References

1. Aguayo JB, Blackband SJ, Schoeniger J, Mattingly M, Hintermann M (1986) NMR imaging of a single cell. Nature 322:190–191
2. Aguayo JB, Cheng HM (1987) Magnetic resonance microscopy of ocular tissues. Med Sci Res 15:1059–1060
3. Aguayo JB, Blackband SJ, Wehrle JP, Glickson JD, Mattingly M (1987) NMR microscopic studies of eyes and tumors with histological correlation. Ann NY Acad Sci 508:399–413
4. Cheng HM, Aguayo JB, Moore GJ, Mattingly M (1991) Analysis of diabetic cataractogenesis using chemical-shift microscopy. Magn Reson Med 17:62–68
5. Cheng HM (1991) Magnetic resonance imaging of the human eye in vivo. Optom Vis Sci 68:976–980
6. Chien D, Buxton RB, Kwong KK, Brady TJ, Rosen BR (1988) Quantitative diffusion imaging in the human brain. 7th annual meeting. Soc Magn Reson Med Abstr: 218
7. Farmer THR, Johnson GA, Cofer GP, Maronpot RR, Dixon D, Hedlund LW (1989) Implanted coil MR microscopy of renal pathology. Magn Reson Med 10:310–323
8. Stejskal EO, Tanner JE (1965) Spin diffusion measurements: spin echos in the presence of a time-dependent field gradient. J Chem Phys 42:288–292

9. Verkman AS, Wong KR (1987) Proton nuclear magnetic resonance measurement of diffusional water permeability in suspended renal proximal tubules. Biophys J 51:717–723
10. Yoshida A, Kwong KK, Chang C, Karino K, Iwasaki T, Buzney S, McMeel JW, Cheng HM (1992) Magnetic resonance microscopy of rabbit eyes. Vision Res 32: 37–40

High-Field ^{31}P Magnetic Resonance Studies for Evaluating Kidney Viability

S. Pomer[1] and W.E. Hull[2]

Introduction

The application of ^{31}P magnetic resonance (MR) to provide a noninvasive evaluation of renal phosphorus metabolism is well established [7, 13, 14]. In a variety of ^{31}P MR studies of excised or perfused kidneys initial attention has been given to determining the time course of nucleotide (NTP + NDP) depletion (sum of all nucleoside 5′-tri- and diphosphates, primarily ATP and ADP) and intrarenal acidosis during cold and warm ischemia [9]. Within a few minutes depletion of intrarenal NTP was noted, and a rapid onset of acidosis was observed. Previous investigations using low-field ^{31}P MR of the isolated perfused kidney demonstrated sequential changes in phosphate components during ischemia [9]. However, the information provided about the components of different signal groups was insufficient; the renal ^{31}P MR spectra obtained at field strengths lower than 4.7 T did not allow adequate sensitivity and resolution for many of the individual phosphate metabolites, for example, detection of phosphocreatine (PCr), separation of NTP and NDP, resolution of phosphomonoesters (PME) into phosphocholine (PC) and phosphoethanolamine (PE) and of diesters (PDE) into glycerophosphocholine (GPC) and glycerophosphoethanolamine (GPE). Through the use of very high field strengths (>7 T) these problems can be solved, at least for ex vivo studies.

The purpose of this study was: (a) to use high-field ^{31}P MR spectroscopy (202 MHz at 11.7 T) of the rat kidney ex vivo to quantitate renal pH and NTP and other phosphate metabolite levels during and after an ischemic insult, and (b) to determine the influence of a postischemic infusion of renoprotective components such as the calcium antagonist diltiazem on the metabolite levels in cold-stored kidneys and to correlate these levels with the recovery of renal function following transplantation.

---

[1] Department of Urology, University Clinic, Im Neuenheimer Feld 110, D-69120 Heidelberg, FRG
[2] Central Spectroscopy Department, German Cancer Research Center, Im Neuenheimer Feld 280, D-69120 Heidelberg, FRG

In the past a major obstacle for the determination of specific metabolic parameters that may be helpful in establishing biochemically meaningful viability predictors has been the inability at low magnetic fields to detect residual NTP in cold-stored kidneys or to separate clearly the large inorganic phosphate (P_i) signal from the unresolved bands denoted as PME and PDE. The simple ratio of signal intensities PME/P_i, proposed as a crude viability parameter [1–3], cannot be measured with great confidence at low fields (1–2.4 T) and is meaningless if kidneys are flushed with a phosphate-containing preservation solution such as EuroCollins or the University of Wisconsin preservative. At 11.7 T the ^{31}P MR spectra provide a wealth of detailed quantitative information which we have used to characterize more accurately ischemic renal injury and recovery in excised rat kidneys [5]. Furthermore, we correlated the MR parameters with subsequent renal function posttransplantation in nephrectomized recipients, as determined by serum creatinine levels, and tested the applicability of crude predictors such as PME/P_i or more biochemically meaningful quantities such as the residual level of nucleotides (NTP + NDP) in the kidney.

Our basic research using high-field NMR also has clinical significance since current technical developments in magnet design have resulted in the commercial production of a 7-T/21-cm system, and possibly a 9.4-T/21-cm system in the near future, which could be used for MR assessment of human kidneys prior to transplantation.

Materials and Methods

MR Spectroscopy. All ^{31}P MR spectra of rat kidneys ex vivo were obtained at 202 MHz (11.7 T) using a Bruker AM-500 spectrometer and a 10-mm multinuclear probe head. The sample temperature was regulated at 4°C throughout the measurements. A 2-mm capillary containing a solution of methylene diphosphonate in D_2O was also introduced into the sample tube to provide field/frequency lock and a chemical shift and integration reference. Data acquisition began 5–6 min after kidney excision. Inverse-gated ^{1}H decoupling (1 W) was used, and the repetition time (T_{rep}) was 1 s with a 45° radiofrequency excitation pulse. The broad signal component from membrane phospholipids was suppressed used a DANTE pulse sequence [6, 12] for presaturation. In the initial phase of cold storage successive spectra were acquired for 5 or 10 min; longer acquisitions were used at later times. Before Fourier transformation the spectrum resolution was enhanced using a Lorentz-Gauss lineshape transformation. The spectra were baseline corrected and digitally integrated using the standard Bruker software. Signal integrals were corrected for the effects of partial saturation (20%–30% for PME, P_i, and PDE) caused by T_{rep} = 1 s (T_{rep} = 4 s gave fully relaxed spectra). The corrected signal integrals were converted to concentrations (µmol/g), which were summed to give the total concentration of phosphate groups detected (total phosphate, ca. 25 µmol/g). Individual metabolite levels were then derived from the corresponding signal integral and expressed as percentage of total phosphate (%P). For NTP + NDP the total integral for the overlapping NTPγ and NDPβ signals was used. The mean intracellular pH was derived from the chemical shift position of the P_i resonance using our own calibration curve determined for reference solutions at 4°C.

Ischemia Studies. Male Wistar rats weighing ca. 150 g with 0.6–0.8 g kidneys were used. Operations were performed on a microsurgery workplace with microscope and thermostatted plate (38°C). After anesthesia was introduced with ether, the abdomen was opened. The left kidney was mobilized until it was attached at the hilum only. For the experiments involving unflushed kidneys the renal pedicle was then clamped and transected, and the kidney was transferred within 10 s to a beaker containing ice-cold EuroCollins-citrate buffer (EuroCollins with 55 m*M* citrate instead of phosphate buffer; Fresenius, Bad Homburg, FRG). After chilling the kidney was transferred to a precooled 10-mm MR sample tube containing 1.5 ml cold EuroCollins-citrate buffer. The sample tube was then placed in the MR probe head within the magnetic field, and measurements were carried out as described above. The effects of warm ischemia were studied by clamping the renal vasculature in situ for a defined time period before excision. Recovery from ischemia was studied by releasing the clamp and waiting various times before excision.

For the experiments involving flushed kidneys the procedure was as follows. After mobilization of the kidney the aorta was mobilized at three sites: above the left renal artery, below the left renal artery, and just above the aortic bifurcation. The right renal artery was then ligated, and the aorta was clamped above the bifurcation and below the left renal artery. A plastic cannula was introduced and tied into the aorta between these clamps, and the upper clamp was then removed. The aorta was tied above the left renal artery, and the left renal vein was opened. Chilled EuroCollins-citrate preservation solution was then flushed through the kidney using a syringe attached to the plastic cannula. In all cases 8 ml solution was introduced over a period of 2 min. The kidney was then rapidly excised and chilled for MR measurements as described above.

Transplantation Studies. Prior to ischemia kidneys were pretreated in vivo ca. 30 min before organ excision by 15-min systemic intravenous infusion of 0.5 ml of the following solutions: (a) standard saline (0.9% NaCl), (b) the calcium antagonist diltiazem, 25 µg/kg (Dilzem, Goedecke, Freiburg, FRG), (c) 50 m*M* ATP $MgCl_2$, ca. 150 µmol/kg. Following 0 or 30 min warm ischemia, kidneys were then flushed with phosphate-free EuroCollins-citrate preservation solution and excised for cold storage as described above. During cold storage a ^{31}P MR spectrum was recorded for 15 min at 4°C at a time between 2 and 6 h postexcision. Except during MR measurements the kidneys were stored on ice until transplantation into a nephrectomized recipient rat ca. 8 h post-excision. As a measure of renal function, serum creatinine levels at 48 h posttransplantation were determined.

Results

In general, high-field ^{31}P MR allowed detailed quantitative analysis of several individual phosphate metabolites. Figure 1 shows a typical 202 MHz ^{31}P MR spectrum of an untreated, excised rat kidney after 8.5 min at 4°C.

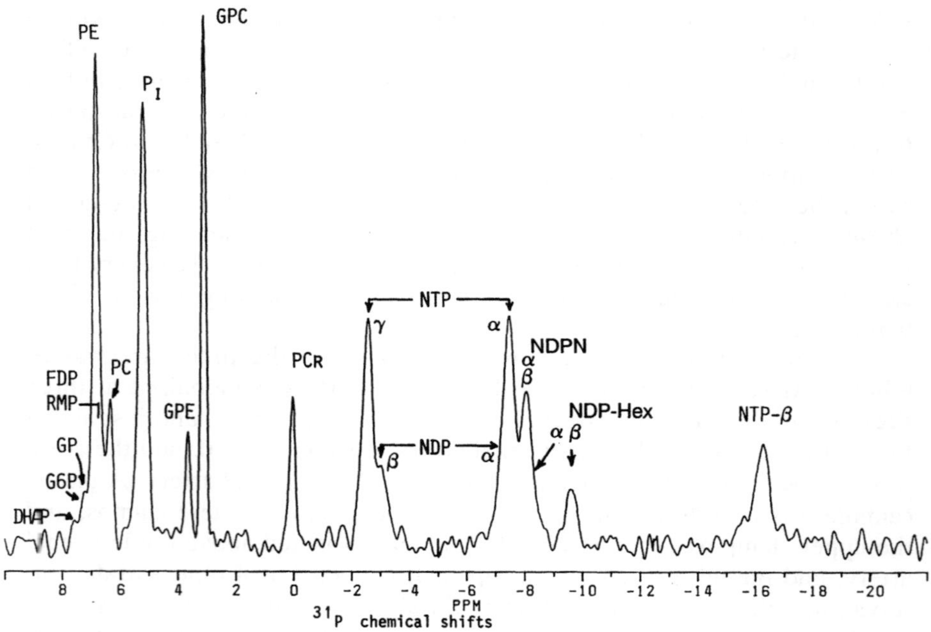

Fig. 1. The ^{1}H-decoupled 202-MHz ^{31}P MR spectrum of an intact, untreated rat kidney measured ex vivo at 4°C, 6–11 min after excision. The chemical shift scale (*horizontal axis*) gives the resonance frequencies of individual peaks (ppm) relative to PCr (1 ppm = 202 Hz). The average intracellular pH was 7.32 as determined from the chemical shift of P_i. The abbreviations used for assigned signals are: *PE*, phosphoethanolamine (includes phosphoserine, if present); *PC*, phosphocholine; P_i, inorganic phosphate (may include glucose 1-phosphate, if present); *GPE*, glycero-3-phosphoethanolamine; *GPC*, glycero-3-phosphocholine; *PCr*, phosphocreatine; *NTP*, all nucleoside 5′-triphosphates (α, β, γ phosphorus atoms); *NDP*, all nucleoside 5′-diphosphates (α, β); *NDPN*, all nucleoside (5′) diphospho (5′)nucleosides (e.g., α,β phosphorus in NAD); *NDP-Hex*, all nucleoside (5′)diphospho(5′)nucleosides (e.g., α, β phosphorus in NAD); *NDP-Hex*, all nucleoside(5′)diphospho(1)hexoses (e.g., α, β phosphorus in UDP glucose. Expected positions for signals from some other metabolites are also marked: *DHAP*, dihydroxyacetone phosphate; *G6P*, glucose 6-phosphate; *GP*, glycerol 3-phosphate; *RMP*, ribose 5′-monophosphates in mononucleotides; *FDP*, fructose 1,6-bisphosphate

The signals were assigned on the basis of their chemical shifts (pH dependent) according to literature data and our own measurements of reference compounds at 4°C as a function of pH. The nucleotide signals are broad because of phosphorus-phosphorus spin coupling and field-dependent linebroadening due to protonation equilibria and magnesium exchange at the phosphate sites. Resonance bands for NTP and NDP can be resolved, but individual signals for specific nucleotides (i.e. ATP, UTP, CTP, GTP) cannot be resolved so that NTP always represents the sum of all four species. Similarly, the signal labeled NDPN contains the α, β phosphorus signals for all nicotine-adenine dinucleotides (oxidized and reduced forms), and NDP-Hex corresponds to two signals representing the α, β phosphates groups of all UDP hexoses such as UDP-Glc, UDP-GlcNAc, and UDP-Gal. When appropriate processing parameters are used, the individual signals for P_i, PCr, PE, PC, GPC, and GPE are well resolved as a result of the high magnetic field strength and the ^{1}H decoupling applied. The PE peak may also contain a contribution from ribose monophosphate, especially in the form of NMP during prolonged ischemia, as well as fructose, 1,6-bisphosphate if present. Additionally, weak signals probably representing small amounts of glycerol 3-phosphate, glucose 6-phosphate, and dihydroxyacetone phosphate are often observed. Generally, the detection sensitivity for such spectra (10–20 min acquisition) is ca. 0.5% of total phosphate, or ca. 0.1 μmol/g.

Cold ischemia showed biphasic behavior in intracellular pH (Fig. 2) and nucleotide loss (Figs. 3, 4). Without EuroCollins-citrate flush the initial pH and NTP level were close to their nominal in vivo values (pH 7.25, NTPγ + NDPβ = ca. 12%, or 3 μmol/g). In the first 2 h there was a rapid decay of pH

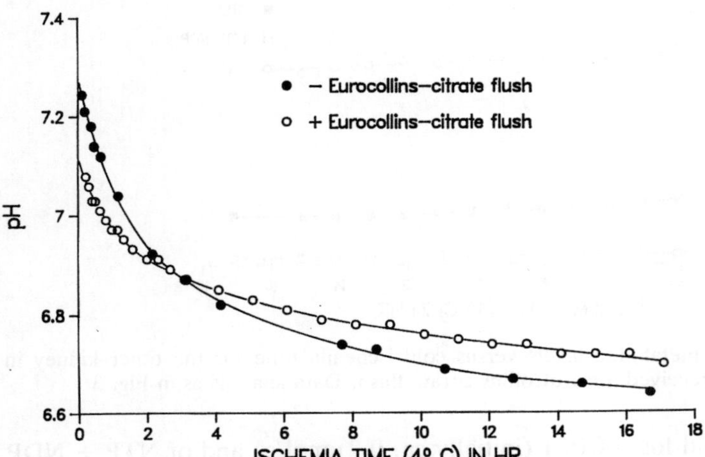

Fig. 2. Kinetics of intracellular pH versus cold ischemia time at 4°C, as determined by successive ^{31}P MR spectra from two rat kidneys: one flushed with EuroCollins-citrate preservation solution (*open circles*) and one without flush (*solid circles*)

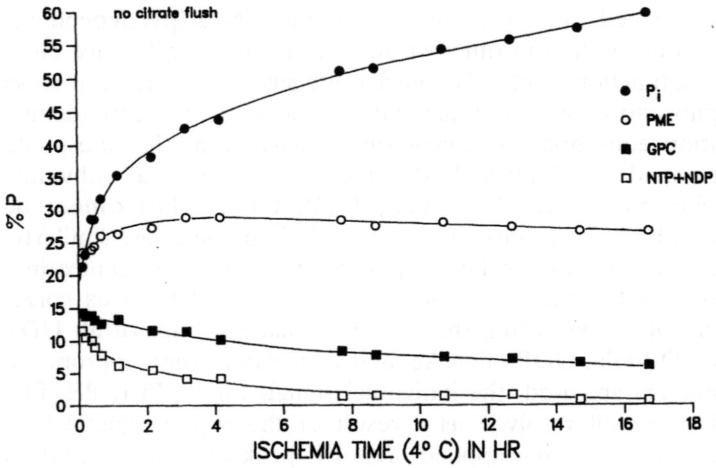

Fig. 3. Phosphate metabolite levels (^{31}P MR signal integrals as percentage of total MR-detected phosphate signals) versus cold ischemia storage time for one of the kidneys of Fig. 2 (without EuroCollins-citrate flush). For NTP + NDP the total integral of the NTPγ + NDPβ signals at ca. -3 ppm was used. For PME the integral over all signals from 6 to 8 ppm was used. The data represent relative molar concentrations

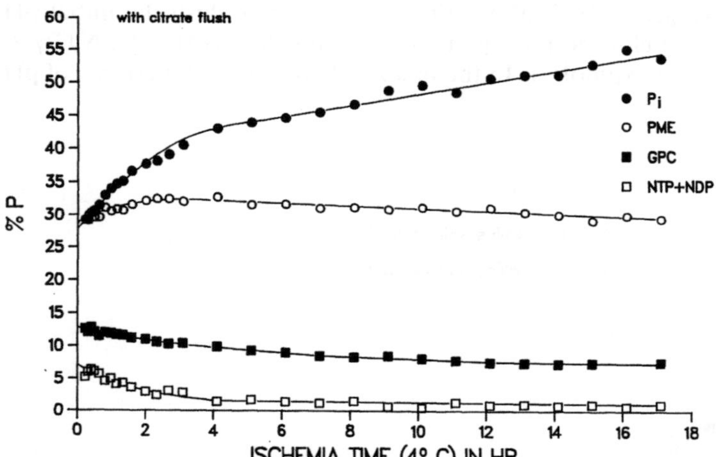

Fig. 4. Phosphate metabolite levels versus cold ischemia time for the other kidney in Fig. 2, that which received a EuroCollins-citrate flush. Data analysis as in Fig. 3

to 6.9 and a rapid loss of PCr (initially ca. 0.7 μmol/g) and of NTP + NDP to give NMP + 2 P_i (exponential time constant 1.4 h at 4°C). This was followed by a slower phase in which the pH, total PME (specifically NMP), GPC, and nucleotides decreased slowly with a time constant of 5–10 h.

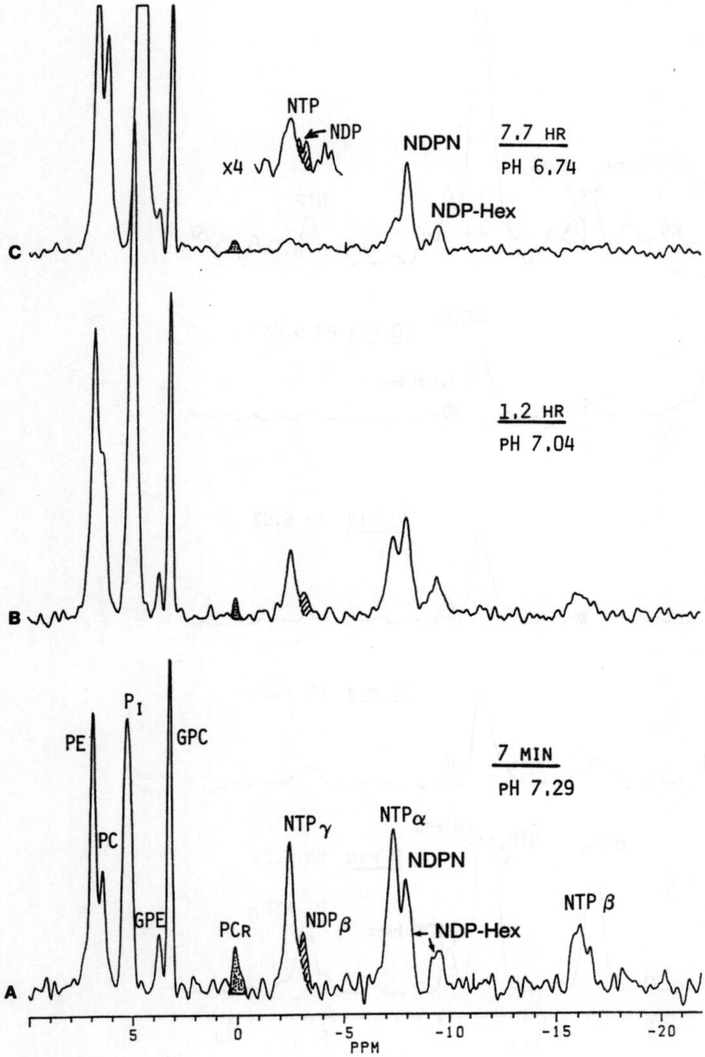

Fig. 5. The 202-MHz ^{31}P MR spectra of a single untreated rat kidney ex vivo after 7 min (**A**), 1.2 h (**B**), and 7.7 h (**C**) at 4°C (no EuroCollins flush). Spectra are labeled with the intracellular pH determined from the P_i chemical shift

NDPN remained stable at a level of ca. 0.35 μmol/g over 18 h of cold storage while NDP-Hex decreased to about 50% of its initial level (ca. 0.75 μmol/g). Representative spectra shown in Fig. 5 illustrate these effects and show that PCr, NTP, and NDP can still be detected even after long storage times. Figure 6 shows that the pattern of metabolite changes during warm ischemia in situ is very similar to that observed in cold storage, but the rate of change

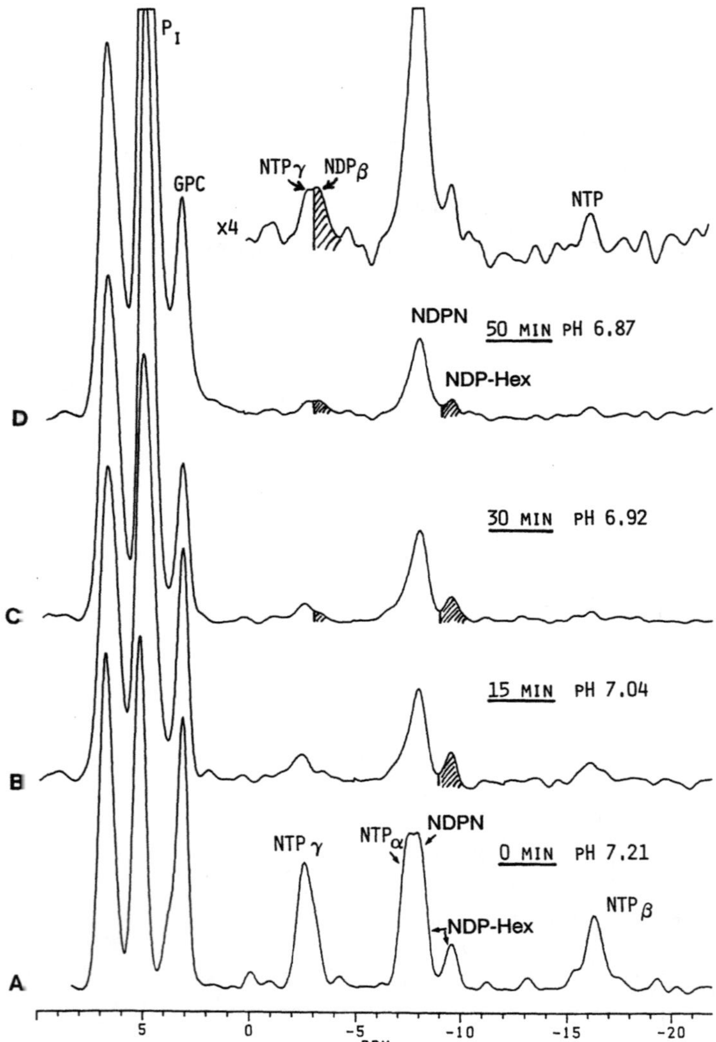

Fig. 6. The 202-MHz ^{31}P MR spectra measured at 4°C within 10 min after excision of rat kidneys subjected to 0 min (**A**), 15 min (**B**), 30 min (**C**), and 50 min (**D**) warm ischemia in situ (no EuroCollins flush). The resolution in these spectra is poorer than in Fig. 5 due to the choice of processing parameters that enhance sensitivity for detecting nucleotide signals

is of course much faster (30 min warm ischemia is about equal to 8 h of cold storage).

With EuroCollins-citrate flush the initially measured pH (7.1) and nucleotide levels (6%) were lowered due to the brief warm ischemia introduced by the flushing procedure, and the initial rapid kinetic phase of Fig. 3

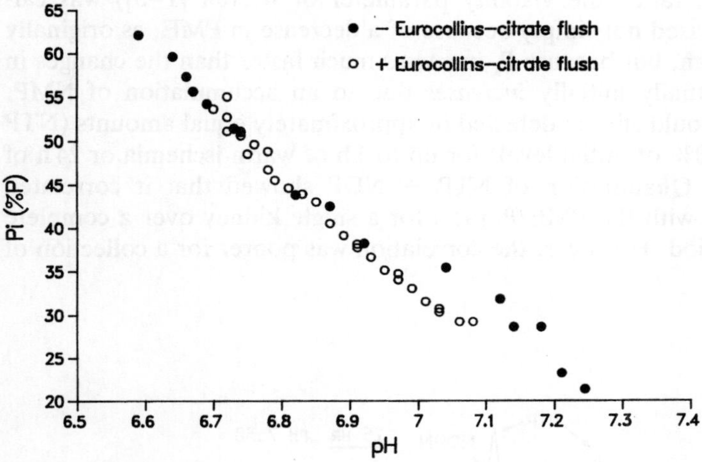

Fig. 7. The approximately linear correlation of P_i concentration versus pH is shown for the two kidneys of Fig. 2 ($r = -0.98$ for all data)

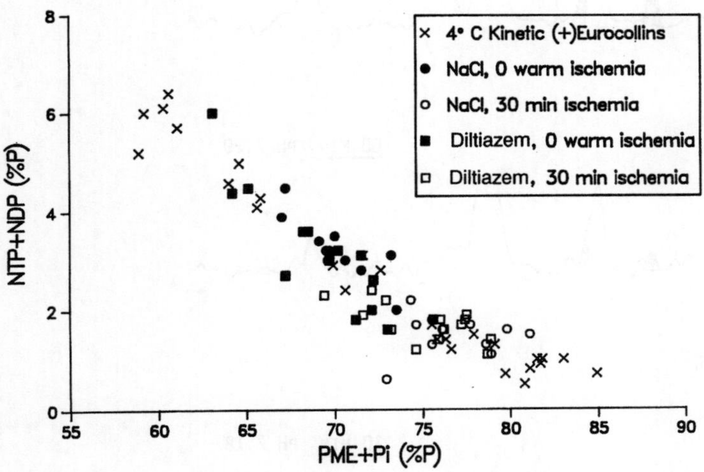

Fig. 8. Correlation plot for NTP + NDP versus PME + P_i for kidneys that received various pretreatments and that were flushed with EuroCollins-citrate before excision. ×, data from a complete time course for the reference kidney (Fig. 4) used for the analyses of Figs. 11 and 12. Each of the other symbols represents a single measurement made on each of the cold-stored kidneys used in the transplantation study

is missing in Fig. 4. EuroCollins-citrate had little effect on metabolite levels observed after 8 h of cold ischemia but did slow the long-term decay of pH (Fig. 2). The use of phosphate-free EuroCollins allowed the direct detection and quantitation of intracellular P_i. There was a good linear correlation between P_i concentration and pH over 18 h of cold storage (Fig. 7).

The PME/P_i ratio (the viability parameter of Bretan [1–3]) was calculated. It decreased not simply because of a decrease in PME, as originally thought by Bretan, but because P_i *increases* much faster than the changes in PME, which actually initially *increases* due to an accumulation of NMP. NTP and NDP could still be detected in approximately equal amounts (NTP + NDP = ca. 10% of initial level) for up to 1 h of warm ischemia or 24 h of storage at 4°C. Quantitation of NTP + NDP showed that it correlated moderately well with the PME/P_i ratio for a single kidney over a complete cold storage period. However, the correlation was poorer for a collection of

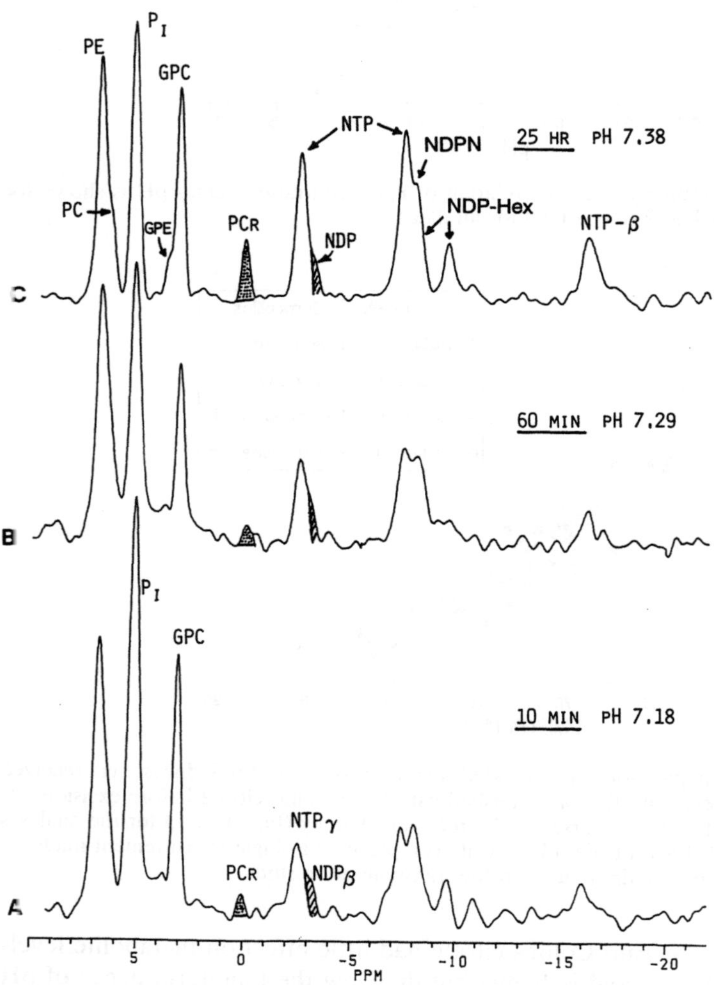

Fig. 9. The 202-MHz ^{31}P MR spectra measured at 4°C ca. 8 min after excision for three different kidneys subjected to 40 min warm ischemia in situ followed by 10 min (**A**), 60 min (**B**), and 24 h (**C**) reperfusion in vivo before excision

kidneys receiving various pretreatments. As expected there was a significantly better correlation between NTP + NDP and PME + P_i, independent of kidney treatment (Fig. 8).

Figure 9 shows ^{31}P MR spectra obtained ex vivo from individual kidneys that were allowed to recover in vivo from warm ischemia (clamped in situ). There was a biphasic behavior in the recovery of nucleotides. Within 10 min after circulation was restored there was ca. 50% recovery of NTP while PCr remained at a low, barely detectable level. After 60 min intracellular pH had returned to 7.3, but there was little further increase in NTP or PCr. Full recovery of NTP and PCr required several hours, but it was not practical to observe this directly with our intraoperative model. Spectra measured after ca. 24 h of recovery were very similar to those obtained from fresh kidneys, and NTP and PCr levels were even somewhat higher than those found in kidneys that received no ischemic insult. Remarkably, even after a 2-h warm ischemia full metabolic recovery (based on the levels of detectable phosphates) was obtained after 24 h for a kidney that would not be expected to have adequate functional recovery based on transplantation experience.

Intravenous pretreatment of animals with diltiazem or saline (control) had no significant effect on metabolite levels in the kidney as detected by ^{31}P MR. During the period of ischemia the decline in tissue NTP levels

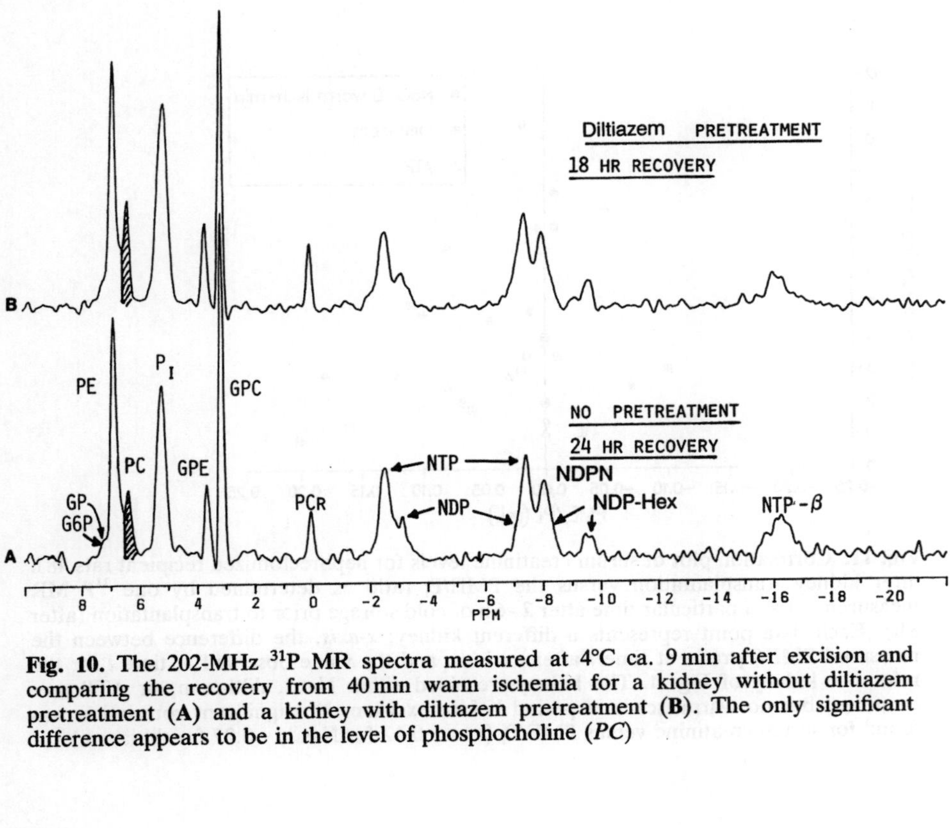

Fig. 10. The 202-MHz ^{31}P MR spectra measured at 4°C ca. 9 min after excision and comparing the recovery from 40 min warm ischemia for a kidney without diltiazem pretreatment (**A**) and a kidney with diltiazem pretreatment (**B**). The only significant difference appears to be in the level of phosphocholine (*PC*)

was identical to the decline observed in the control rats. The recovery of metabolite levels following reperfusion in vivo was not significantly affected by pretreatment (Fig. 10), except possibly for some influence on the level of PC. Thus, the infusion of diltiazem had no detectable influence on nucleotide levels during ischemia and did not appear to improve recovery in the 1st h or after 24 h.

For cold-stored kidneys involved in the transplantation study the single MR measurement of each kidney could not, for practical reasons, be made at exactly the same time after excision. Therefore, the measured metabolite concentrations at a particular time point were compared to those obtained at the same time point for a reference kidney whose complete cold ischemia kinetic profile had been determined (Fig. 4). The *differences* in these values (transplanted kidney's relative to reference values) were used for subsequent analysis.

For transplanted kidneys that were not subjected to a warm ischemia period and were stored for ca. 8 h after flushing with EuroCollins, the creatinine values obtained after 2 days did not correlate well with the Bretan predictor PME/P_i (Fig. 11) but did show a reasonable correlation with the parameter NTP + NDP (Fig. 12), as measured during cold storage, especially for those kidneys that received only saline as a pretreatment

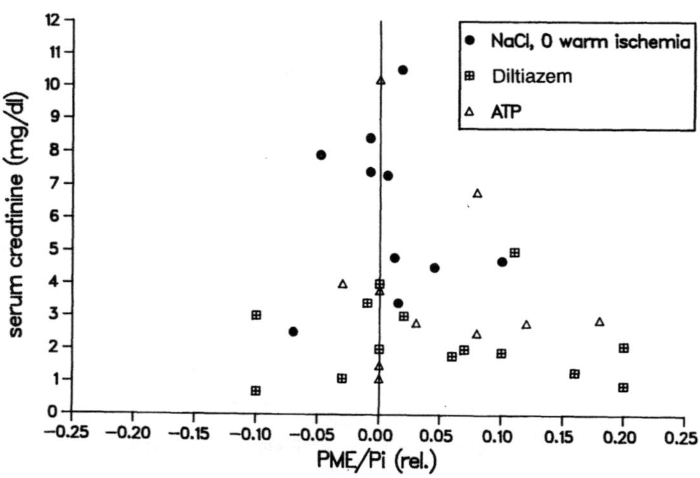

Fig. 11. Correlation plot of serum creatinine levels for nephrectomized recipient rats 48 h after kidney transplantation versus the PME/P_i ratio as determined by one ^{31}P MR measurement at a particular time after 2–6 h of cold storage prior to transplantation (after 8 h). Each data point represents a different kidney; *x-axis*, the difference between the measured PME/P_i ratio at time *t* after excision and the value observed at time *t* for the reference kidney of Fig. 4. The kidneys received either NaCl, diltiazem, or ATP pretreatment but no warm ischemia period before excision. No significant correlation was found for serum creatinine versus PME/P_i ($r = -0.12$ for NaCl, -0.27 for diltiazem)

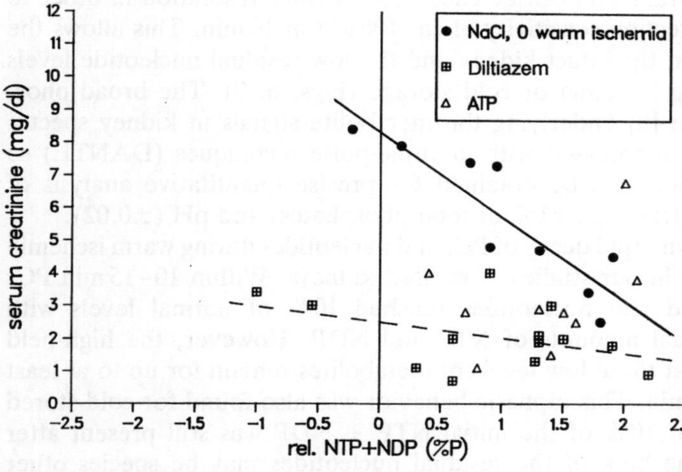

Fig. 12. Correlation plot for serum creatinine versus NTP + NDP (relative to the values for the reference kidney). The kidneys and treatments were those of Fig. 11. *Solid line*, the linear regression line ($r = -0.89$) obtained for the data representing the NaCl (control) pretreatment; *dashed line*, the regression line ($r = -0.42$) obtained for the diltiazem pretreatment

($r = -0.89$, $\beta = -2.57 \pm 0.49$). Higher nucleotide levels during storage correlated with lower serum creatinine values (better function). After transplantation, function (serum creatinine, but not NTP + NDP levels) depended on the type of preischemia treatment given. Diltiazem treatment gave the best functional results, NaCl controls had the worst results, while ATP pretreatment gave intermediate results that did not correlate with the measured NTP + NDP levels. The improvement afforded by diltiazem versus saline increased as the relative NTP + NDP level monitored in cold storage decreased.

For kidneys subjected to 30-min warm ischemia before excision, there was no clear correlation between function (serum creatinine) after transplantation and NMR-detected metabolite levels (either PME/P_i or NTP + NDP, data not shown). There was more scatter in the results, but the trend in pretreatment efficacy of diltiazem > ATP > saline remained.

Discussion

The high-field (202 MHz, 11.7 T) ^1H-decoupled ^{31}P MR of rat kidneys ex vivo demonstrates high spectral dispersion with resolution of many individual metabolites, particularly in the PME and PDE regions, when appropriate processing parameters are chosen (Figs. 1, 5, 10). Alternatively, data processing parameters can be chosen to sacrifice resolution in order to achieve a high detection sensitivity of ca. 100 μ*M* in 10 min. This allows the detection of PCr in the intact kidney and the low residual nucleotide levels still present during ischemia or cold storage (Figs. 6, 9). The broad phospholipid resonance [8] underlying the metabolite signals in kidney spectra can be effectively suppressed with multiple-pulse techniques (DANTE) so that a good baseline can be obtained for precise quantitative analysis of metabolite concentrations (±1% of total phosphates) and pH (±0.02).

The well-known rapid decay of PCr and nucleotides during warm ischemia was also observed in our studies of excised kidneys. Within 10–15 min PCr nearly disappeared and nucleotides reached 10% of normal levels with approximately equal amounts of NTP and NDP. However, the high-field studies showed that these low levels of metabolites remain for up to at least 1 h of warm ischemia. This biphasic behavior was also found for cold-stored kidneys, where ca. 10% of the initial NTP + NDP was still present after 24 h. However, the bulk of the residual nucleotides may be species other than ATP or ADP. As has been reported in earlier studies [10, 11], recovery from ischemia results in a biphasic increase in nucleotide levels (ca. 50% recovery in 10 min). We have found that PCr recovers only during the second, slow phase and requires several hours.

The fact that excellent recovery of phosphate metabolite levels occurred even after a 2-h warm ischemia period suggests that such *metabolic* recovery may occur even when *functional* recovery is incomplete. The calcium antagonist diltiazem had no influence on the kinetics of phosphate metabolites during or after ischemia but did have a clearly positive influence on functional recovery of transplanted kidneys, especially for the kidneys that were more depleted in nucleotides. Thus, the calcium antagonist diltiazem may assist recovery by preventing a strong influx of calcium upon reperfusion when the ATP reserves required for pumping out calcium are depleted. Since renal ischemic injury may primarily involve cell membrane damage, especially during reperfusion, it may be useful to examine more closely the effects of ischemia and reperfusion on the levels of the phospholipid metabolites PC, PE, GPC, and GPE. For example, there has been some indication of a correlation between the duration of ischemia and the PME/PDE ratio measured 21 days postischemia [4]. However, the signals of the individual phospholipid metabolites can be adequately resolved only at high magnetic field strengths.

During cold storage NTP + NDP levels dropped within 8 h to about 10% of nominal values. During this time the total amount of PME increased in proportion by about the amount of NTP + NDP lost, and with resolution enhancement an increase in a signal in the PME region consistent with ribose 5′-monophosphates was detected. Thus, there was an *increase* in PME due to the conversion of NTP and NDP to NMP + P_i in this first phase of cold ischemia, which is not reflected well in Bretan's PME/P_i ratio (*decreases*). In the second phase of cold ischemia total PME decreased slowly, parallel with decreases in pH, GPC, and NTP + NDP. Thus, the exhausting of NMP reserves may be an important factor in long-term kidney viability. Here, again, more detailed studies of the PME signal region are of interest.

In our transplantation studies it was found that Bretan's PME/P_i parameter, measured during cold storage, did not correlate as well as the residual NTP + NDP concentration did with post-transplantation function (creatinine excretion). Although PME includes NMP and P_i increases as nucleotides are degraded, the PME/P_i ratio has no direct biochemical significance in the sense that NTP + NDP concentration has. In view of the limited quality of the low-field ^{31}P MR that was obtainable from intact canine or human kidneys in a clinical setting at the time and the inability to detect residual nucleotides in such spectra, the PME/P_i ratio was virtually the only practical parameter that could be *estimated* from the spectra. A rationale that this parameter can provide a crude estimate of kidney viability for clinical practise can be found in the crude correlation that we observed between PME/P_i and NTP + NDP.

Based on our experiments it would be desirable to use directly the NTP + NDP parameter (as a percentage) of total detected phosphates or possibly the complementary parameter PME + P_i to judge viability. However, these parameters can be measured only when the preservation solution does not contain phosphate. It may in fact be better to use a ratio such as (NTP + NDP)/(NDPN + NDP-hex) which involves only nucleotides and can be measured even in the presence of P_i in the perfusate. To this end we are currently developing methods for studying intact, cold-stored canine kidneys at 7 T using a magnet with 15-cm bore, and we find that it is possible to detect and quantitate residual nucleotides and intracellular pH (based on the chemical shift of PE) even after 24-h storage and even with phosphate buffer present. With the introduction of commercial 7-T/21-cm magnets, which can accommodate cold-stored human kidneys, the future looks bright for the clinical application of high-field MR for judging kidney viability.

References

1. Bretan PN, Vigneron DB, Hricak H, Juenemann K-P, Williams RD, Tanagho EA, James TL (1986) Assessment of renal preservation by phosphorus-31 magnetic resonance spectroscopy: in vivo normothermic blood perfusion. J Urol 136:1356–1359
2. Bretan PN Jr, Vigneron DB, James TL, Williams RD, Hricak H, Juenemann K-P, Yen TSB, Tanagho EA (1986) Assessment of renal viability by phosphorus-31 magnetic resonance spectroscopy. J Urol 135:866–871
3. Bretan PN Jr, Baldwin N, Novick AC et al. (1989) Pretransplant assessment of renal viability by phosphorus-31 magnetic resonance spectroscopy. Transplantation 48:48–53
4. Gonzales-Mendez R, Nabsath DC, Stevenson DK, Goms ML, Schulz TE, Jardetzky O (1989) Renal injury and regeneration in adult rats after total ischemia, MRS correlations. Magn Reson Med Biol 2:21–32

5. Hull WE, Pomer S, Vogt P (1987) High-field ^{31}P-NMR analysis of phosphate metabolites in excised rat kidney: assessment of preservation for transplantation. Abstr Soc Magn Reson Med 6:281
6. Morris GA, Freeman RA (1978) Selective excitation in Fourier transform NMR. J Magn Reson 29:433–462
7. Ross B, Freeman D, Chan L (1986) Contributions of nuclear magnetic resonance to renal biochemistry. Kidney Int 29:131–141
8. Seelig J (1978) P-31 NMR and the head group structure of phospholipids in membranes. Biochim Biophys Acta 515:105–140
9. Sehr PA, Bore PJ, Papatheofanis J, Radda GK (1979) Non-destructive measurement of metabolites and tissue pH in the kidney by ^{31}P nuclear magnetic resonance. Br J Exp Pathol 60:632–641
10. Siegel NJ, Avison MJ, Reilly HF, Alger JR, Shulman RG (1983) Enhanced recovery of renal ATP with post-ischemic infusion of ATP-$MgCl_2$ determined by ^{31}P-NMR. Am J Physiol 245:F530–534
11. Stromski ME, Cooper K, Thulin G, Gaudlo GM, Siegal NJ, Shulman RG (1986) Chemical and functional correlates of post-ischemic renal ATP levels. Proc Natl Acad Sci USA 83:6142–6145
12. Thoma WJ, Snyder JA, Pearson GA (1989) Broad resonance-edited ^{31}P spectra obtained with a single radiofrequency channel. NMR Biomed 2:112–114
13. Weiner MW, Adam WR (1985) MRS for evaluation of renal function. Semin Urol III:34–42
14. Wong GG, Ross BD (1983) Application of phosphorus nuclear magnetic resonance to problems of renal physiology and metabolism. Miner Electrolyte Metab 9:282–289

Clinical Magnetic Renonance

Overview of Clinical ^{31}P Magnetic Resonance Spectroscopy Investigations and Applications for Renal Transplant Viability Assessment

P.N. Bretan, Jr.[1]

Phosphorus-31 magnetic resonance spectroscopy (^{31}P MRS) is a nondestructive and noninvasive technique in which intracellular phosphorus metabolites are used as viability parameters for assessment of cadaveric renal viability pretransplantation and to delineate ischemic metabolic pathways for formulation of better organ preservation maneuvers. This overview summarizes the development of ^{31}P MRS for the assessment of renal transplant viability in the setting of acute ischemia, with its current clinical applications. In addition, the use of ^{31}P MRS for preservation studies and in situ assessment of renal transplant metabolism using surface coils is discussed.

Pretransplant Ex Situ Assessment of Renal Viability Using MRS

The initial experiments (Fig. 1, top) in the development of ^{31}P MRS renal viability parameters used isolated ex vivo rat kidneys that were subjected to varying degrees of warm and cold ischemia. Specific MRS parameters were correlated with previously established electron microscopic (EM) viability parameters and parallel studies of postischemic renal function. It was found that a time-dependent decay of the phosphomonoester/inorganic phosphate (PME/Pi) ratios existed over an observed 70 h of simple cold storage. These correlations of the ^{31}P MRS ratios help predict subsequent renal function and viability (Fig. 1, bottom). Ratios of 0.30–0.45 corresponded with acute tubular necrosis (ATN), while ratios above this range were associated with completely normal renal function, and ratios less than these limits were associated with irreversible ischemic renal injury and death in the rat model [1]. These preliminary contentions that the PME/Pi ratios were strong viability parameters were supported in a subsequent study with canine kidneys monitored by MRS of biopsy specimens [2]. These ex situ studies provided the motivation to verify these parameters in the in vivo, in situ dog model (Fig. 2). These subsequent canine MRS studies showed a graded

[1] Harbor-UCLA Transplant Service, University of California at Los Angeles, School of Medicine, Los Angeles, CA90033, USA

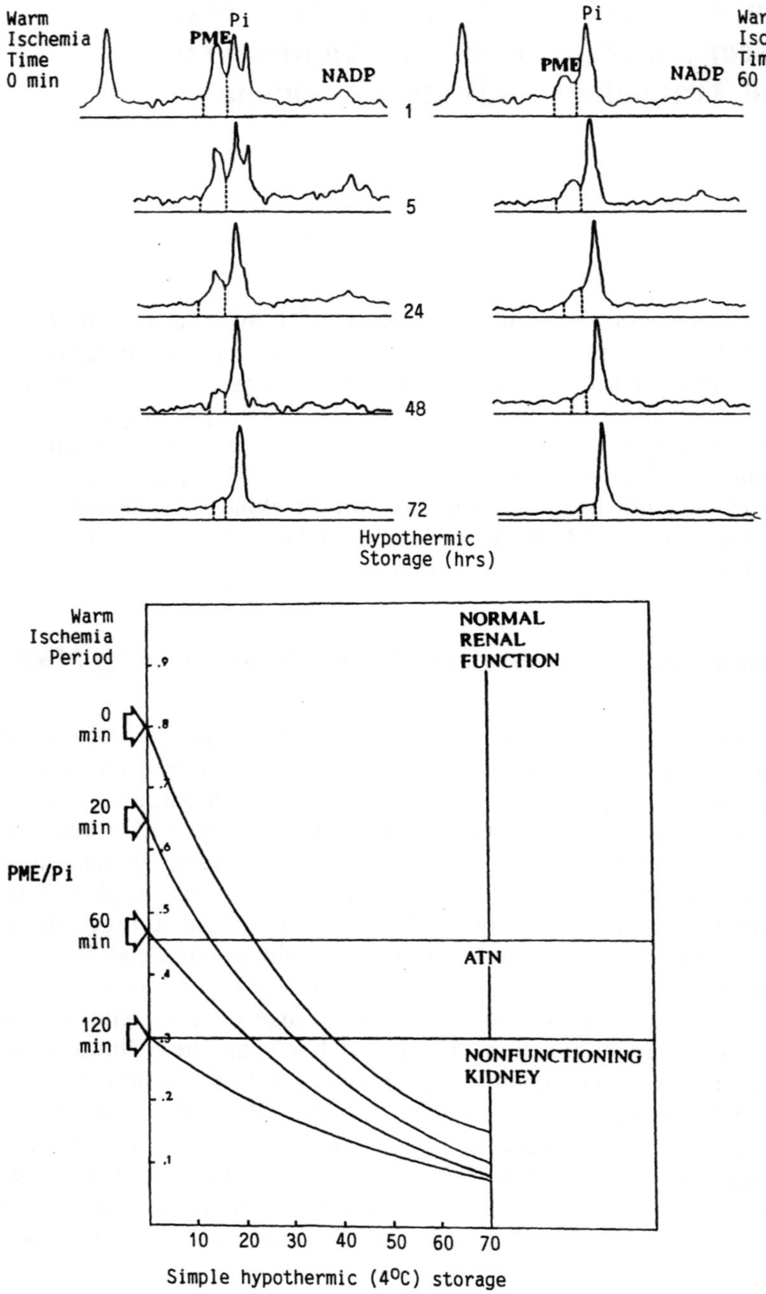

Fig. 1. Serial ^{31}P MRS spectrum (*top*) and cumulative data (*bottom*). MRS-derived phosphomonoester (*PME*) over inorganic phosphate (*Pi*) ratios. Time-dependent decay over 70 h of hypothermic storage and 120 min of warm ischemia of PME/Pi enables renal viability assessments. (From [1])

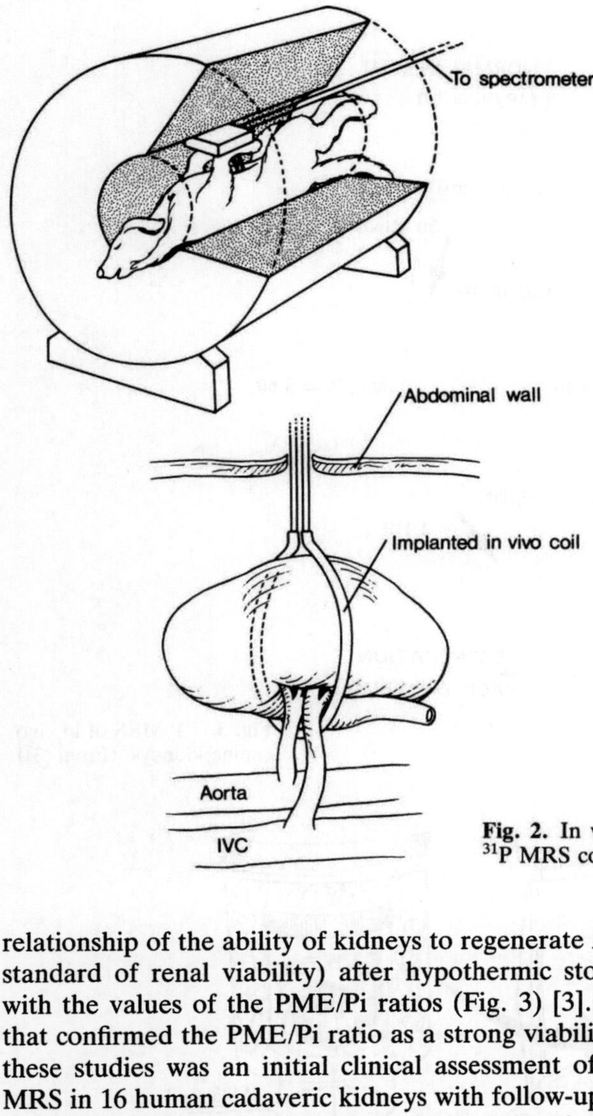

Fig. 2. In vivo canine kidney; implanted ^{31}P MRS coil. (From [3])

relationship of the ability of kidneys to regenerate ATP (considered the gold standard of renal viability) after hypothermic storage and warm ischemia with the values of the PME/Pi ratios (Fig. 3) [3]. It was these correlations that confirmed the PME/Pi ratio as a strong viability parameter. Included in these studies was an initial clinical assessment of renal viability using ^{31}P MRS in 16 human cadaveric kidneys with follow-up in the recipient patients. Thirteen kidneys had high ratios (0.85 ± 0.08) and were associated with excellent renal function in the recipient patients as noted by urinary output (1589 ± 693 ml on posttransplant day 3) and corresponding mean drop in serum creatinine (8.6 ± 5.2 to 3.4 ± 2.1 mg/dl). Thus, these 13 kidneys were verified to be viable. Three other kidneys had depressed PME/Pi ratios of 0.37 ± 0.2 associated with azotemia (serum creatinine 9.1 ± 3.0 mg/dl) and oliguria (<200 ml/day).

The largest clinical series using ^{31}P MRS for pretransplant assessment of renal viability involved 40 cadaveric kidneys (Fig. 4). Pretransplant MRS

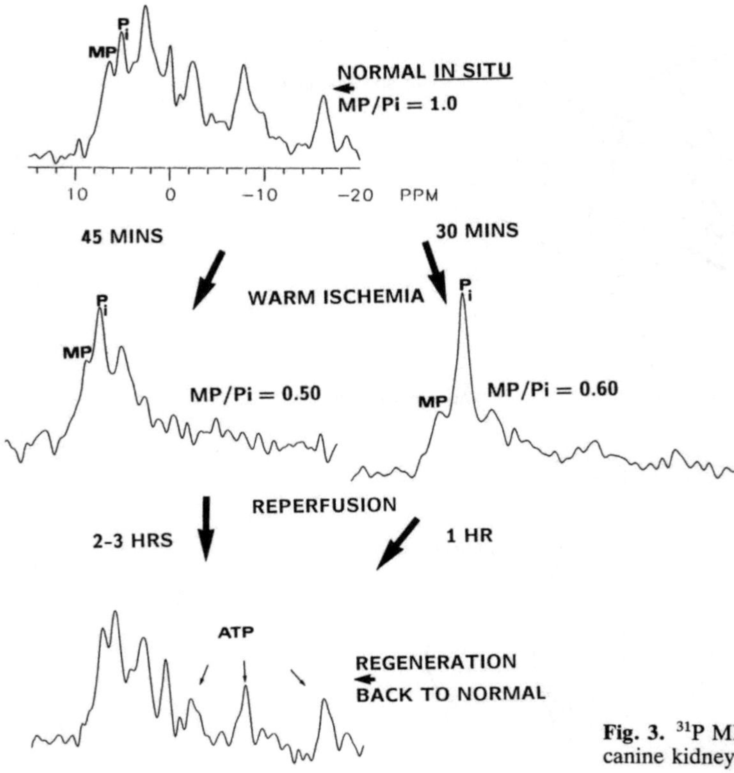

Fig. 3. ^{31}P MRS of in vivo canine kidneys. (From [3])

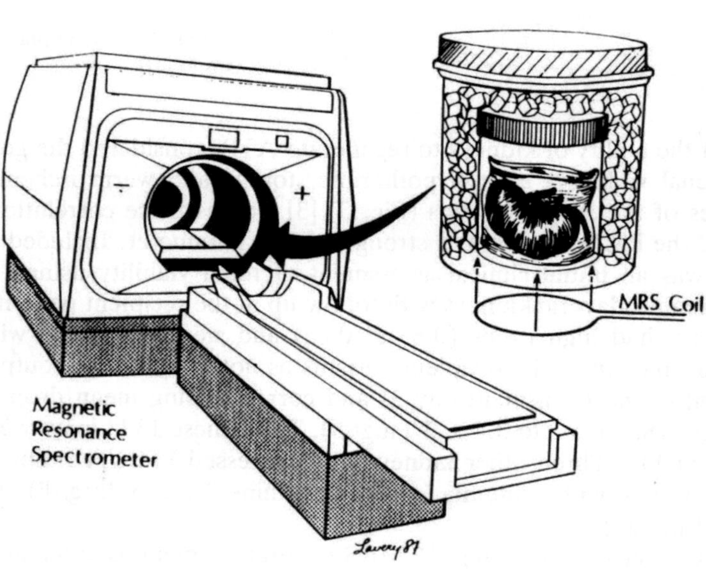

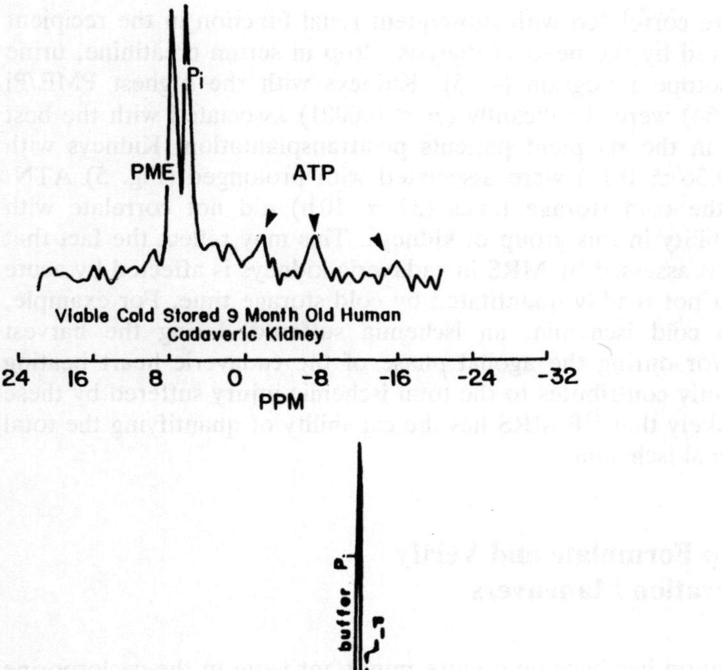

Fig. 5. Phosphorus-31 spectrum (24.3 MHz) of an ex situ in vivo viable pediatric kidney (*top*) and adult kidney (*bottom*). Phosphomonoester (*PME*), inorganic phosphate (*Pi*), and ATP peaks defined. (Modified from [4] with permission)

←

Fig. 4. Cadaveric kidney in sterile hypothermic storage container with external MRS coil and cold storage container within superconducting magnet bore. (From [4] with permission)

parameters were correlated with subsequent renal function in the recipient patients measured by the need of dialysis, drop in serum creatinine, urine output, and isotope renogram [4, 5]. Kidneys with the highest PME/Pi ratios (1.66–0.54) were significantly ($p < 0.0001$) associated with the best renal function in the recipient patients posttransplantation. Kidneys with lower ratios (0.36 ± 0.10) were associated with prolonged (Fig. 5) ATN. Interestingly, the cold storage times (37 ± 10 h) did not correlate with subsequent viability in this group of kidneys. This may reflect the fact that renal viability as assessed by MRS in cadaveric kidneys is affected by acute global ischemia not readily quantitated by cold storage time. For example, in addition to cold ischemia, an ischemia suffered during the harvest procedure and/or during the agonal phase of the cadaveric heart beating donor significantly contributes to the total ischemic injury suffered by these kidneys. It is likely that ^{31}P MRS has the capability of quantifying the total acute global renal ischemia.

MRS To Help Formulate and Verify Renal Preservation Maneuvers

Organ preservation has become a more important issue in the cyclosporine era of renal transplantation. With the use of cyclosporine, a synergistic nephrotoxic effect is noted when a significant ischemic insult is sustained by a donor kidney prior to transplantation [6–8]. In addition, there is increasing evidence that moderate ischemic injury predisposes renal allografts to severe rejection and diminished survival [6, 9–11]. For nonrenal organs, such as hearts, liver, and pancreas, preservation is even more important. An increase in the safe preservation window of these organs would extend the geographical limits for procurement and would facilitate sharing among transplant centers, both regionally and nationally.

A number of new intracellular renal flush solutions have been found more efficacious than Collins 2 (C-2) solution in extending organ viability during simple cold storage. However, the mechanism of action of these solutions remains poorly understood. We used MRS to delineate better the underlying intracellular mechanisms involved in preservation. To accomplish this we studied a modified simple hypothermic intracellular (340 mosm/kg) flush solution (PB-2). The development of PB-2 solution is based on the ability of some of its individual components to minimize ischemic adenine nucleotide (AN) catabolism and the endothelial microvascular damage also referred to as post-reperfusion injury [12] (Table 1). These preliminary results in ten canine autorenal transplants showed a significant ($p < 0.02$) improvement in renal recovery and viability (recipient posttransplant inulin clearance and survival) after 50 h of cold storage compared to ten canine kidneys similarly preserved using conventional C-2 flush solution.

Table 1. Composition of different intracellular renal flush solutions (g/l; from [12])

	PB-2 Flush	Collins 2 Flush	Sacks 2 Flush	Belzer Perfusate	UW-1 Flush	PB-3 Flush
KH_2PO_4	2.05	2.05	4.16	3.4	3.4	2.05
$K_2HPO_43H_2O$	9.70	9.70	9.70	–	–	9.70
KCL	1.12	1.12	–	–	–	1.12
KHCO3	–	–	2.30	–	–	2.30
Mannitol	25.0	–	37.5	–	–	25.0
Glucose	–	25	–	1.5	–	–
$MgSO_47H_2O$	3.70 or	7.38	–	8	1.2	–
$MgCl_2$ (mEq/ml)	0.4	–	2	–	–	0.4
Adenosine	1.0	–	–	1.3	1.34	1.0
Sodium glutathione	–	–	–	17.5	0.92	–
Albumin	–	–	–	5.3	–	–
$NaHCO_3$	0.84	0.84	1.26	–	–	0.84
Allopurinol	–	–	–	0.113	0.113	0.100
Verapamil	–	–	–	–	–	0.007
K^+ lactobionate	–	–	–	–	39.8	–
Raffinose	–	–	–	–	17.8	–
Hydroxyethyl starch	–	–	–	–	50	–
Osmolality (mosm/kg)	340	320	430	300	320–330	340
pH	7.25	7.00	7.00	7.10	7.40	7.30–7.40

Standard methods in which to assess renal viability were used, such as high performance liquid chromatography (HPLC) studies. HPLC showed a significant ($p < 0.01$) loss of AN using C-2, while PB-2 was associated with regeneration of AN within 45 min of reperfusion (Fig. 6). Similarly, ^{31}P MRS showed greater high-energy phosphate metabolites (PME or NADP, $p < 0.001$) after 50 h of cold storage using PB-2 (Fig. 7a) compared to C-2 (Fig. 7b). EM revealed normal microcapillary morphology for the PB-2 group; however, moderate vascular red and white blood cell clumping were observed in the C-2 group.

Characterization of PB-2 solution's basic preservation mechanism by HPLC, ^{31}P MRS, and EM studies indicates that PB-2 solution enhances renal preservation both by diminution of reperfusion injury and by decreasing the loss of intracellular high-energy metabolites that are necessary for viability. Thus, ^{31}P MRS has been used to delineate better the metabolic changes that are affected by acute ischemia and to establish improved preservation maneuvers in order to diminish these effects in conjunction with standard HPLC and EM methods. The development of new organ preservation flush PB-2 solution has been shown to decrease the post-transplant reperfusion injury by keeping PME/Pi ratios high during prolonged cold storage [12].

In addition, PB-3 solution (addition of verapamil and allopurinol to PB-2; see Table 1) has been shown to have a significant salutary effect on preservation in isolated mitochondrial oxidative phosphorylation during 72 h

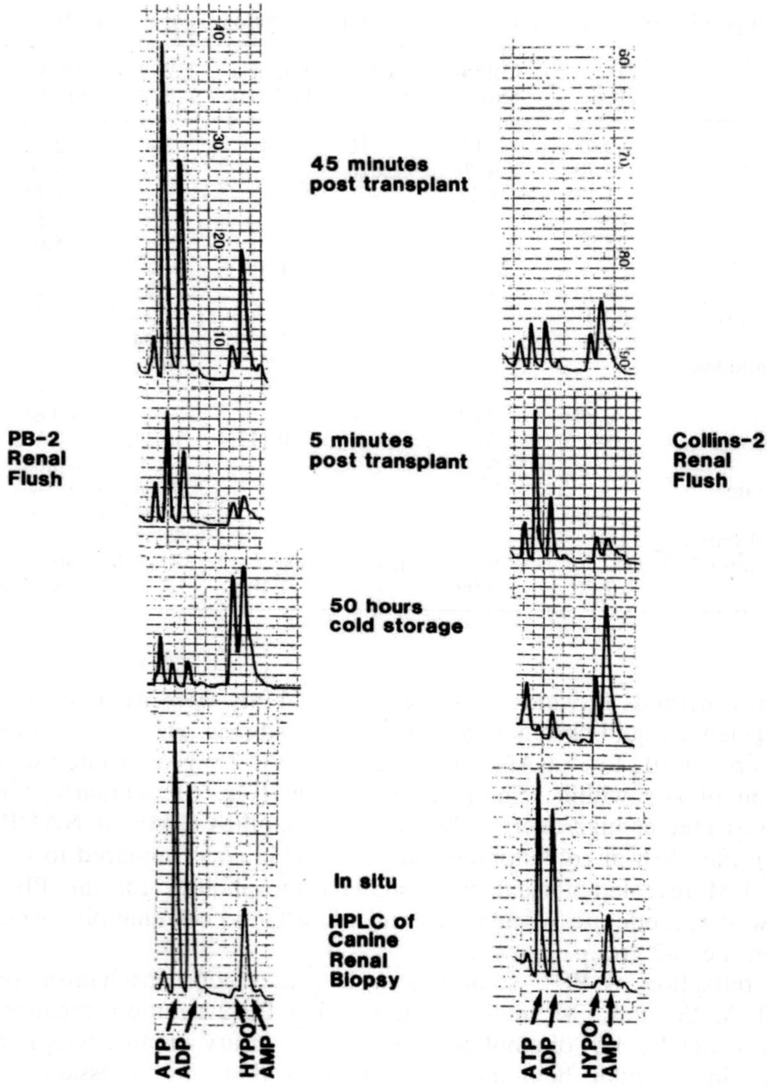

Fig. 6. HPLC of canine kidneys. Note superior regeneration of AN by PB-2 during 5–45 min of reperfusion, while further deterioration of AN during this time interval in the C-2 group is characteristic of the reperfusion injury. (From [12])

of hypothermic storage compared to C-2 solution (P.N. Bretan et al., unpublished). It is hoped that MRS use in these preservation settings will help to ascertain many of the intracellular metabolic pathways that are altered during acute ischemia of cold storage and how these pathways interact.

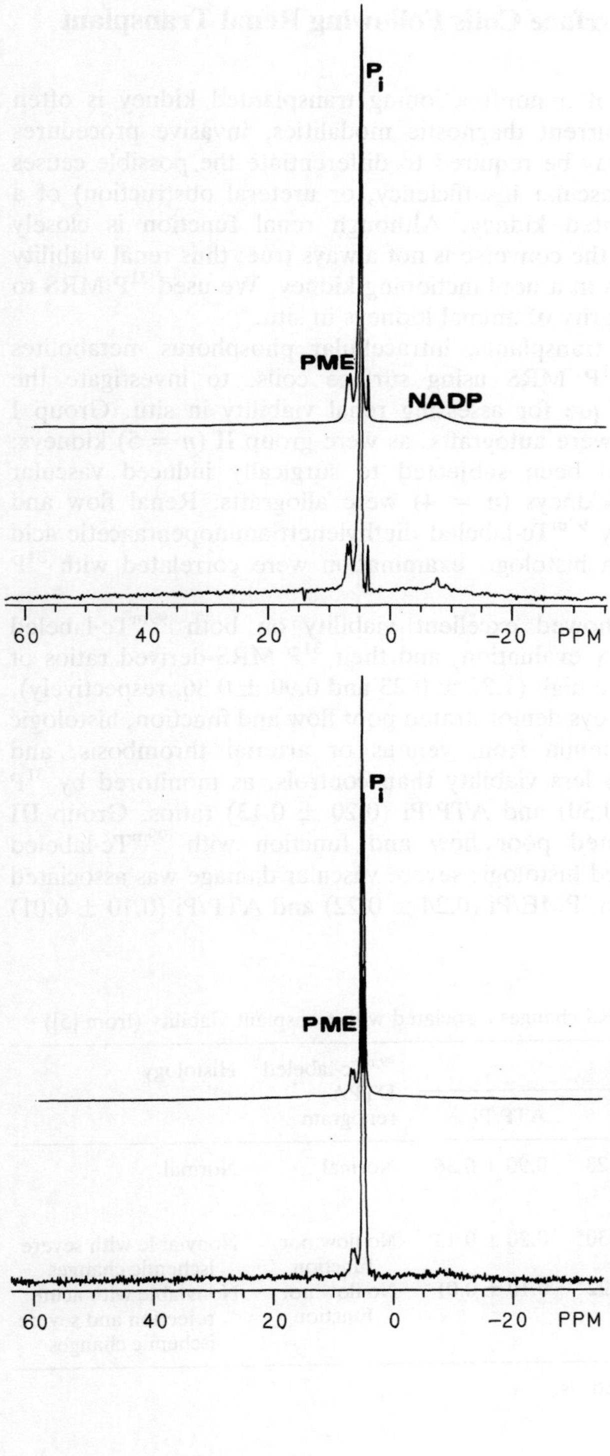

Fig. 7. Representative ^{31}P MR spectra from whole canine kidneys after 50 h of cold storage, just prior to autorenal transplantation. PB-2 flush (*top*) is associated with greater preservation of PME (predominantly AMP for kidneys [15]) and NADP compared to C-2 flush (*bottom*). (From [12])

In Situ MRS Using Surface Coils Following Renal Transplant

Noninvasive evaluation of a nonfunctioning transplanted kidney is often difficult [13]. Despite current diagnostic modalities, invasive procedures (such as renal biopsy) may be required to differentiate the possible causes (i.e., ATN, rejection, vascular insufficiency, or ureteral obstruction) of a nonfunctioning transplanted kidney. Although renal function is closely associated with viability, the converse is not always true; thus renal viability remains difficult to assess in a nonfunctioning kidney. We used ^{31}P MRS to assess the metabolic integrity of animal kidneys in situ.

In 14 canine renal transplants, intracellular phosphorus metabolites were monitored with ^{31}P MRS using surface coils, to investigate the usefulness of this technique for assessing renal viability in situ. Group I control kidneys (n = 5) were autografts, as were group II (n = 5) kidneys; however, the latter had been subjected to surgically induced vascular thrombosis. Group III kidneys (n = 4) were allografts. Renal flow and function, as measured by ^{99m}Tc-labeled diethylenetriaminopentaacetic acid (DPTA) and findings on histologic examination were correlated with ^{31}P MRS spectra (Table 2).

Group I kidneys showed excellent viability on both ^{99m}Tc-labeled DPTA studies and biopsy evaluation, and their ^{31}P MRS-derived ratios of PME/Pi and ATP/Pi were high (1.32 ± 0.23 and 0.90 ± 0.36, respectively). In contrast, group II kidneys demonstrated poor flow and function, histologic evidence of severe ischemia from venous or arterial thrombosis, and significantly ($p < 0.005$) less viability than controls, as monitored by ^{31}P MRS PME/Pi (0.58 ± 0.30) and ATP/Pi (0.20 ± 0.13) ratios. Group III kidneys also demonstrated poor flow and function with ^{99m}Tc-labeled DTPA, and the associated histologic severe vascular damage was associated with accelerated rejection. PME/Pi (0.24 ± 0.22) and ATP/Pi (0.10 ± 0.01)

Table 2. Canine renal ^{31}P MRS changes associated with transplant viability (from [5])

Group	n	^{31}P MRS		^{99m}Tc-labeled DTPA renogram	Histology
		PME/Pi	ATP/Pi		
I Autografts, viable controls	5	1.32 ± 0.23	0.90 ± 0.36	Normal	Normal
II Autografts, nonviable	5	0.58 ± 0.30*	0.20 ± 0.13*	No flow nor function	Nonviable with severe ischemic changes
III Allografts, nonviable	4	0.24 ± 0.22*	0.10 ± 0.01*	No flow nor function	Nonviable with acute rejection and severe ischemic changes

*$p < 0.0005$ compared to controls.

ratios were also significantly ($p < 0.005$) less than those in controls, reflecting nonviability. Thus the ^{31}P MRS-derived PME/Pi and ATP/Pi ratios enable noninvasive assessment of renal viability; however, this technique could not differentiate severe ischemic injuries from those caused by severe acute rejection.

Currently, these and many other in situ studies concentrate on improvement of localization of the MRS signal from the renal cortex [14]. The in situ dog model using various surface coils could provide a setting in which to study such localization techniques. It is hoped that these developments will improve specificity and sensitivity for these current MRS viability parameters in the clinical setting.

Conclusion

^{31}P MRS has been successfully used noninvasively and sterilely to assess renal viability in a postharvest, pretransplant setting during cold storage. The use of PME/Pi ratios give a quantitative analysis of viability, while ATP/Pi enables a very specific qualitative assessment of viability. In addition, ^{31}P MRS has been used to help delineate specific ischemic metabolic pathways associated with cold storage, helping to formulate and verify improved organ preservation maneuvers such as the use of the new PB flush solutions. With this more rational approaches to organ flush preservation development can evolve. Finally, ^{31}P-MRS using specially designed surface coils has potential use as a posttransplant in situ study to differentiate ischemic injury in the renal allograft from other causes of dysfunction postoperatively, adding a useful tool in the clinical management of the transplant patient postoperatively.

Acknowledgement. I wish to express special thanks to Juana Avina for preparation of this manuscript.

References

1. Bretan PN, Vigneron DB, James TL, Williams RD, Hricak H, Jeunemann P, Yen B, Tanagho EA (1986) Assessment of renal viability by phosphorous-31 magnetic resonance spectroscopy. J Urol 135:866–871
2. Bretan PN, Vigneron DB, Hricak H, Juenemann KP, Tanagho EA, James TL (1986) Assessment of renal preservation by phosphorous-31 magnetic resonance spectroscopy – in vivo normothermic blood perfusion. J Urol 136:1356–1359
3. Bretan PN, Vigneron DB, Hricak H, Price D, Collins GM, Tanagho EA, James TL (1987) Assessment of clinical renal preservation by ^{31}P-MRS. J Urol 137:146–150
4. Bretan PN, Baldwin N, Novick AC, Majors A, Easley K, NG TC, Stowe N, Streem S, Steinmuller D, Rehm P, Go R (1989) Clinical experience with pretansplant

assessment of renal viability by phosphorus-31 magnetic resonance spectroscopy (^{31}P-MRS) in 40 recipient patients. Transplantation Proc 21:1266–1267
5. Bretan PN, Baldwin N, Novick AC et al. (1989) Pretransplant assessment of renal viability by ^{31}P-MRS. Transplantation 48:48–53
6. Opelz G (1986) Multicenter impact of cyclosporine and cadaver kidney graft survival. Prog Allergy 38:329–345
7. Novick AC, Ho-Hsieh H, Steinmuller D, Streem SB et al. (1986) Detrimental effect of cyclosporine on initial function of cadaver renal allografts following extended preservation – results of a randomized prospective study. Transl 42:154
8. Erturk E, Bretan PN, Pestana J et al. (1989) The relative effects of aztreonam vs gentamycin on nephrotoxicity induced by warm ischemia in the presence and absence of cyclosporine. Transplant Proc 21:932–935
9. Keller H, Fischer G, Kirste G, Wilms H (1989) ATN influence on renal transplant function. Transplant Proc 21:1282
10. Maessen JG, Greve J, Buurman WA et al. (1989) Sensitivity of ischemically damaged kidneys to inflammatory reaction. Transplant Proc 21:1261
11. Bretan PN (1989) Extracorporeal renal preservation. In: Novick AC (eds) Stewart's textbook of operative urology. Williams and Wilkins, Baltimore
12. Bretan PN, Baldwin N, Martinez A, Stowe N, Scarpa A, Easley K, Erturk E, Jackson C, Pestana J, Novick AC (1991) Improved renal transplant preservation using PB-2 flush solution – characterization of intracellular mechanisms by renal clearance, HPLC, ^{31}P-MRS and EM Studies. Urol Res 19:73–80
13. Hattner RD, Engelstad BL, Dae MW (1984) Radionuclide evaluation of renal transplants. In: Freeman LM, Weissman HS (eds) Nuclear medicine annual. Raven, New York, pp 319–342
14. Bretan PN, Vigneron DB, Hricak H, Proce DC, Yen TSB, Luo JA, Tanagho EA, James TL (1991) Assessment of renal transplant viability by ^{31}P-MRS – experimental study in canines. Am Surg (in press)
15. Balwin N, Bretan PN, Scarpa A (1988) Predominance of renal phosphomonoester (PME) ^{31}P-MRS peak by AMP – correlative and quantitative analysis using HPLC assay. Proc Soc Magn Reson Med [Works in Progress Suppl]:77

RARE Magnetic Resonance Urography in the Diagnosis of Upper Urinary Tract Abnormalities in Children*

G. Sigmund[1], B. Stoever[1], L.B. Zimmerhackl[2], A. Frankenschmidt[3], E. Nitzsche[4], J.U. Leititis[2], F.E. Struwe[1], and J. Hennig[1]

Patients and Methods

Patients

From September 1989 to November 1990, 55 consecutive children (35 boys, 20 girls) with suspected morphologic pathology of kidneys and urinary tract were prospectively examined by magnetic resonance imaging (MRI). No other patient selection criteria were applied. In 25 of the cases suspicion had been aroused by an abnormal finding on an ultrasound screening examination. Twenty-two patients were examined due to a pathologic finding on ultrasound (US), which had been performed because of urinary tract infection, and/or due to other reasons like multiple malformations with expected kidney involvement (8 cases). Four children were evaluated twice – yielding a total of 59 examinations. These four repeated examinations were scheduled for control of therapy and are not included in this presentation. The patients were between 2 days and 13.2 years old, half of them were under 6 months old.

Thirty-four received an oral sedative 45 min prior to the scan; one child was examined under general anesthesia because a cystoscopy was to follow immediately. Twenty children were given no medication at all (Table 1). All the children had undergone US. In 27 an EU was performed, in 39 a VCUG, and in 41 renal scintigraphy using technetium-99m mercaptoacetyltriglycine (Tc-99m MAG_3) with furosemide intervention. Three kidneys were nonfunctional (scintigraphy: excretory function ≤2% on the respective side), 7 had a very low excretory function (scintigraphy: 5%–14% on the respective side; almost "silent" on EU), and in three renal duplications one of the two segments was nonfunctional. In 20 patients subsequent urinary tract surgery was carried out, six children are scheduled for cystoscopy and/or operation, and one child with related cardiac and other malformations died of urosepsis.

---

* First published in *Pediatric Radiology* (1991) 27:416–420.
Departments of [1]Diagnostic Radiology, [2]Pediatrics, [3]Urology, and [4]Nuclear Medicine, University of Freiburg, D-79104 Freiburg, FRG

At the time of the MR examination, only the provisional sonographic diagnosis was given to the MRI readers. The quality of the images was subjectively classified as: *excellent*, if virtually no motion artifacts and a high signal-to-noise ratio was present; *satisfactory*, if there were motion artifacts and considerable noise, but the images were still diagnostic; *poor*, if images were deemed nondiagnostic.

Methods

The rapid acquisition with relaxation enhancement (RARE) technique uses long echo-trains with different phase encoding for each echo; thereby the number of excitations necessary for a 256^2 matrix is reduced compared to conventional spin-echo sequences. Thus, heavily T2-weighted images are generated in a reasonable acquisition time. A modification of this technique called RARE MR hydrography has a T2 weighting such that only substances with a T2 relaxation time of 500 ms or longer will produce a signal, resulting in aqueous fluid-specific images ("water pictures"). The special advantage of this modification is the even shorter acquisition, with a total scan time of 6.4 s per slice using one excitation (TE = 25 ms, 256^2 matrix, total scan time = 256 × 25 ms = 6400 ms = 6.4 s). Routinely, two averages are used for one RARE MR hydrography slice, with a recovery time of 10 s. Thus, total scan time with TR/TE = 10 s/25 ms is 23 s [6.4 s (first acquisition) + 10 s (recovery) + 6.4 s (second acquisition) = 22.8 s ≈ 23 s]. The RARE sequences including RARE MR hydrography belong to the standard software of our resistive midfield system (Bruker Tomikon R 23 operating at 0.23 Tesla; Bruker Medizintechnik, Karlsruhe, FRG). Recent developments in hardware have also enabled the use of this method on a high-field system [5, 11]. Technical details of RARE and its variants are described elsewhere [5–8]. For the most common clinical applications the technique is referred to as RARE MR myelography [3, 7] or RARE MR urography [2, 4] although RARE MR hydrography (or simply water pictures) is the most comprehensive term.

In this study, RARE MR urograms in coronal, sagittal, and transversal planes were obtained in each case. The smallest fitting coil was used: Adult's knee or head coil for the infants, neck or body coil for the older children. Finally, a coronal T1-weighted conventional spin-echo sequence (7 min/12 slices) was performed; in selected cases sagittal T1-weighted slices were also acquired.

Results

The quality of the images was subjectively classified by two observers based on freedom from motion artifacts and good signal-to-noise ratio (see above). Thus, 56 of the 59 RARE MR urograms and 51 of the T1-weighted sequences

Table 1. Preparation of pediatric patients and image quality ($n = 59$ examinations)

Preparation	Quality of RARE MR urography		Quality of T1-weighted sequence
Oral/rectal sedative[a] 45/20 min prior to scan ($n = 38$)	Excellent	32	25
	Satisfactory	4	8
	Poor	2	5
General anesthesia[b] ($n = 1$)	Excellent	1	1
No medication:			
Feeding in infants: ($n = 9$)	Excellent	8	6
	Satisfactory	1	2
	Poor		1
School-age children: ($n = 11$)	Excellent	9	8
	Satisfactory	1	1
	Poor	1	2

For final diagnosis surgery/cystoscopy (if performed), and/or the combination of all imaging modalities were used as reference.
[a] Oral promazine 3–5 mg/kg bd. wt. (Protactyl, Wyeth, Münster, FRG), in 4 cases chloral hydrate 600 mg per rectum (Chloralhydrat-Rectiole, Dentinox, Berlin, FRG).
[b] Because of subsequent cystoscopy.

Table 2. Correct MRI diagnosis of urinary tract dilatation (number correctly described by MR/total number)

Final diagnosis[a]	Hydronephrosis/ dilated pelvis	Megaureter/ dilated ureter
Pelviureteric obstruction (including 6 on both sides)	19/19	–
Dilatation of ureter and associated collecting system		
Due to vesicoureteric reflux III°–V° (including 2 renal duplications)	10/12	9/12
Due to vesicoureteric reflux II°	–	1/1
Nonrefluxing (including 6 renal duplications)	11/11	11/11
Total	40/42	21/24

[a] Based on surgery, cystoscopy, and/or combination of all other imaging studies.

were regarded as satisfactory or excellent (Table 1). At least one of the two techniques could be evaluated in each patient, allowing a diagnosis to be made in each case.

All 19 hydronephrotic kidneys resulting from pelviureteric obstruction were clearly demonstrated (Table 2, Fig. 1). Twenty-one out of 24 dilated ureters (mildly dilated or megaureter) and 21 out of 23 associated dilated pelvicaliceal systems (mildly dilated with normal renal parenchyma or hydronephrosis) were identified (Table 2, Figs. 2, 3), including those without excretory function. A differentiation could not be made between non-

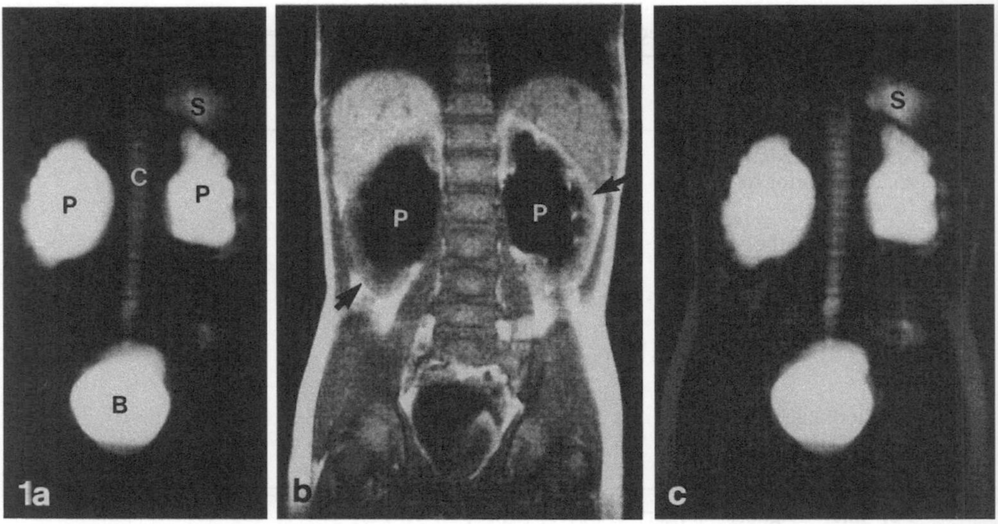

Fig. 1a–c. A 2.5-year-old boy with mild urinary tract infection. **a** Coronal RARE MR urography: Massive dilatation of renal pelvis (*P*) and calices on both sides. *S*, Fluid-filled stomach; *C*, cerebrospinal fluid; *B*, urinary bladder. Ureters are not delineated, i.e., not dilated. **b** Representative T1-weighted slice shows thin parenchyma (→), but no severe atrophy. Urine is depicted *dark* (long T1 = low signal). **c** Algebraic addition of (**a**) and (**b**). Image postprocessing allows for better anatomical-topographic orientation, e.g., to locate the fluid in the stomach (*S*)

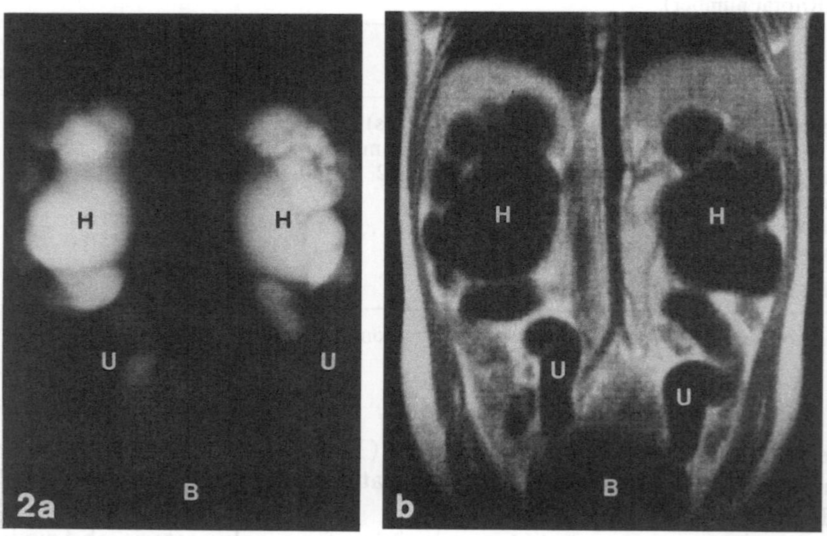

Fig. 2a,b. A 9.9-year-old neglected boy with untreated urethral valves, vesicoureteric reflux V° on both sides, and yet compensated renal insufficiency. **a** Coronal RARE MR urography. **b** Coronal T1-weighted image: bilateral massive hydronephrosis (*H*) with severe parenchymal atrophy and bilateral megaureters (*U*); B, bladder

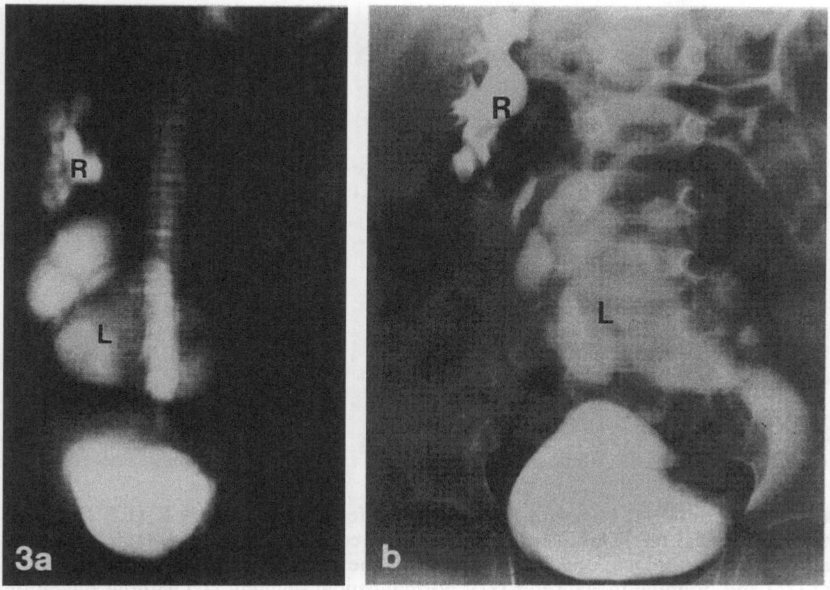

Fig. 3a,b. A 3-year-old boy with crossed dystopia of hydronephrotic "left" kidney (*L*) and associated megaureter (scintigraphy: 15% excretory function). The right kidney (*R*) shows mild dilatation due to pelviureteric obstruction (however, 85% of excretory function). Similar morphologic information of coronal RARE MR urography (**a**) and conventional excretory urography (**b**)

refluxing dilatation and dilatation due to vesicoureteral reflux. In one case with vesicoureteric reflux II° the MR image suggested reflux because of a mild dilatation of the ureter (but not of the renal pelvis). A vesicoureteric reflux I° on the opposite side was missed. Severely reduced excretory function could be suspected due to the loss of corticomedullary differentiation and atrophy of the renal parenchyma on the T1-weighted images (Fig. 2).

Eight out of nine renal duplications were identified (Fig. 4). In the one false negative there was no dilatation of either segment. Size and multiplicity of renal cysts were shown in all eight cases, including even small cystic lesions of Potter I polycystic nephropathy. In one case only the calices were dilated, with the renal pelvis appearing otherwise normal (differential diagnosis: megacalices/pelvicalyceal obstruction). As yet, surgical intervention in this child has been postponed; however, the US and MRI findings are in full agreement. The morphology of one crossed dystopia (Fig. 3) and another dysplasia were correctly described, two children with borderline sonographic findings were correctly identified as normal. Two ureteroceles were detected by RARE MR urography, two smaller ones were missed. Incidentally, a fibrous band forming a horseshoe kidney, and a small urachal cyst were found upon surgery performed because of other reasons. These secondary

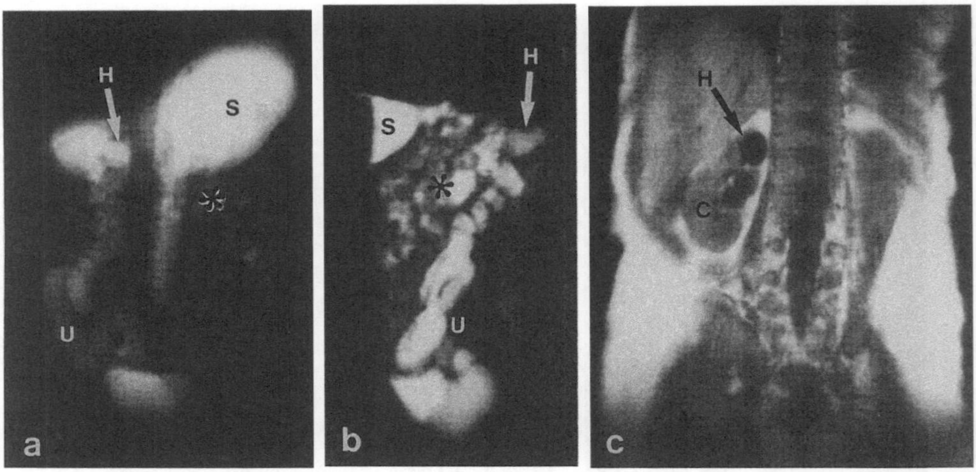

Fig. 4a–c. A 2-month-old boy with urinary tract infection. **a** Coronal RARE MR urography. **b** Right sagittal RARE MR urography – corresponding to a lateral view. **c** Coronal T1-weighted image: renal duplication on the right with hydronephrotic cranial segment (*H*) and draining megaureter (*U*), normal caudal segment (*C*) without dilatation of lower collecting system. Fluid in the gastrointestinal tract (*S*, stomach – with air-fluid level in **b**; *, small bowel), which leads to superposition in **a** but can be ascribed to stomach and small bowel in **b**

findings had been missed preoperatively by all imaging methods. On the other hand, no false positive MRI findings were encountered in this study.

Discussion

The outstanding feature of RARE MR urography is that it rapidly and selectively images fluid in cysts, calices, renal pelvis, dilated ureter, and urinary bladder. In contrast to conventional T1-weighted multislice sequences, *one* 5- to 10-cm-thick slice is sufficient to depict the *entire* collecting system with minimal superposition. Only structures that are filled with fluid can cause superposition, such as the subarachnoid space, the gastrointestinal tract, the gallbladder, the urinary bladder, or the tortuous ureter itself. By choosing other slice orientations or thinner slices these minor superpositions can be ascribed to the respective organ system or be completely faded out. Solid tissues and gas produce no signal at all and therefore cannot cause any superposition. Thus, bowel motion or gas do not interfere with the quality of the RARE urogram. This feature and the visualization of the entire collecting system in one slice represent the main advantages of RARE MR urography over US.

The coronal RARE MR urogram resembles the AP view of an ordinary EU, however, it is independent of excretory function. And again it has the advantage that no bowel gas interferes with quality of the image.

The coronal T1-weighted spin-echo sequence can provide additional information on the renal parenchyma, especially in nondilated segments, and on the corticomedullary differentiation [1]. Furthermore, image post-processing, that is, the addition of a RARE MR urogram to an appropriate T1-weighted slice, allows for better anatomical-topographic orientation.

What are the major limitations? The RARE MR urogram is a static image, so that at present it is not possible to diagnose a vesicoureteric reflux. Only the morphologic changes, i.e., megaureter and hydronephrosis, can be visualized, but they cannot be distinguished from dilatation due to obstruction or other causes. However, the same drawback applies to US, even with additional duplex or color Doppler equipment [9, 10]. Thus, for diagnosing vesicoureteric reflux voiding cystourethrography is still considered to be the gold standard.

Spatial resolution of RARE MR urography is inferior to X-ray urography, so that only a dilated ureter can be clearly demonstrated. Of course, MR is inferior to sonography in terms of costs and time resolution, too. Regarding morphology, the cheaper ultrasound will be sufficient for diagnosing pathologies such as hypoplastic or dystopic kidneys, parenchymal damage, cysts, stones, and ureteroceles, where MRI could not yield additional information. A result that is of major relevance for MRI in infants and young children in general is that there is virtually no need for general anesthesia or deep sedation. For young babies normal feeding often produces sufficient sedation.

Conclusions

RARE MR urography is a new diagnostic tool to examine urinary tract abnormalities in children – including newborns. Besides being independent of excretory function, it requires no contrast medium and employs no ionizing radiation, which is a major benefit, especially in the case of infants. The total scan time of 23 s per slice (6.4 s/single acquisition) is very short.

The main advantage of the method is that in contrast to ultrasound, one slice provides an image of the entire urinary tract without interference caused by bowel motion or gas. By combining RARE MR urography with T1-weighted images, additional information on the renal parenchyma can be obtained in nondilated segments, for example, and anatomical-topographic orientation is facilitated. At present, it is not possible to identify a vesicoureteric reflux, so that VCUG is still mandatory for the differential diagnosis.

However, these preliminary results suggest that MR can replace EU in certain cases such as pelviureteric obstruction, renal duplication, and megaureter and hydronephrosis, if US, VCUG and renal scintigraphy have produced corroborating findings. In this way, one examination with radiation exposure can be eliminated.

References

1. Dietrich RB, Kangarloo H (1986) Kidneys in infants and children: evaluation with MR. Radiology 159:215
2. Friedburg H, Hennig J, Frankenschmidt A (1987) RARE-MR-urography: imaging of the urinary tract with a new fast non-to-mographic MR technique. Radiologe 27:45
3. Friedburg HG, Hennig J, Schumacher M (1987) Clinical application of RARE-MR-myelography as a routine MR examination: Experience with 175 cases. ROFO 146:584
4. Friedburg H, Westenfelder M, Roeren T, Hennig J (1987) MR of the urinary tract in pregnancy. ROFO 147:430
5. Hennig J (1988) Multiecho imaging sequences with low refocusing flip angles. J Magn Reson 78:397
6. Hennig J, Friedburg H (1988) Clinical applications and methodological developments of the RARE technique. Magn Reson Imaging 6:391
7. Hennig J, Friedburg H, Ströbel B (1986) Rapid nontomographic approach to MR-myelography without contrast agents. J Comput Assist Tomogr 10:375
8. Hennig J, Nauerth A, Friedburg H (1986) RARE imaging: a fast imaging method for clinical MR. Magn Reson Med 3:823
9. Jequier S, Paltiel H, Lafortune M (1990) Ureterovesical jets in infants and children: duplex and color Doppler US studies. Radiology 175:349
10. Marshall JL, Johnson ND, DeCampo MP (1990) Vesicoureteric reflux in children: prediction with color Doppler imaging. Work in progress. Radiology 175:355
11. Mulkern RV, Wong STS, Winalski C, Jolesz FA (1990) Contrast manipulation and artifact assessment of 2D- and 3D-RARE sequences. Magn Reson Imaging 8:557

Magnetic Resonance Imaging of Solid Renal Masses

T. Roeren, G.M. Richter, and V. Ihle[1]

In the adult solid renal tumors are of mesenchymal or epithelial origin. The latter occur more frequently, while benign or malignant mesenchymal lesions are rare. Solid masses can also be due to renal involvement in extrarenal disease (metastases, lymphoma) or direct renal infiltration by tumor spread from adjacent tissues. With sonography and computed tomography (CT) differentiation of these lesions is possible only in cases of typical attenuation or sonomorphology, as in lipomas or angiomyolipomas. In all other cases the similar appearance of solid tumors obviates a specific diagnosis. In the early stages of magnetic resonance imaging (MRI) the kidneys became an immediate focus of attention because of the unique ability of MRI to permit distinction between renal cortex and medulla. With time, however, the first enthusiasm sobered and the value of MRI as a method to stage renal tumors was judged more critically. Studies with larger numbers of patients compared MRI to state-of-the-art CT, which includes a contrast-enhanced study, and came to equivocal conclusions as to which method to choose for detection and staging of renal tumors.

Detection of Renal Tumors with MRI

The majority of malignant solid renal masses are renal cell carcinomas (RCC). These account for approximately 2% of all malignant diseases in the adult. In about 1% we find bilateral and in 5% multifocal occurrence. Exact assessment of tumor spread is important for staging, prognosis, and therapeutic approach. CT and sonography allow an accurate pretherapeutic staging in 90% of RCC cases [5]. While a study published by Hricak et al. in 1985 [3] claimed a 100% detection rate of RCC with MRI, later studies, including one by the same group, had overall lower detection rates of 90%–96%, with detection rates below 90% when tumor size was less then 3 cm [1, 4, 9]. One of the two limitations of MRI in distinguishing between

[1] Department of Diagnostic Radiology, University of Heidelberg, Im Neuenheimer Feld 110, D-69120 Heidelberg, FRG

benign renal parenchyma and solid tumor is a spatial resolution of – as of now – not less than 1 cm in vivo. The second and at the time clinically more important limitation is the unpredictable signal intensity of solid tumors.

Staging of Solid Renal Tumors with MRI

Once a tumor is known, MRI has a number of advantages compared to other imaging methods. The most important are that distinction between collateral vessels and enlarged hilar lymph nodes is virtually always possible, and that tumor thrombus in large vessels (see Fig. 1) is more readily detected [2, 4]. In our institution a study of 65 patients who were imaged for suspected renal tumor correctly defined or excluded extension of tumor into the renal vein or the vena cava. Although intravascular signals may at times not be definitely interpretable as thrombus or flow artifact, the use of fast imaging techniques and variable phase encoding should solve this problem. As with CT, however, MRI may not reliably identify tumor growth through the renal capsule, but this disadvantage is of little clinical importance. Another limitation shared with CT is the inability to reliably determine tumor in regional lymph nodes. In some cases even lymph nodes with signal intensities identical to that of the primary tumor proved to be benign at histopathological work-up [1].

Influence of Imaging Parameters on Detection and Staging

Imaging Plane

Due to its multiplanar imaging capability and the possibility of always projecting the partial volume effect in a direction where it least hinders a proper diagnosis, MRI has a definite advantage over other cross-sectional imaging methods. Horizontal tissue interfaces – difficult to image by CT – can be well examined by MRI when choosing the adequate, usually coronal, plane [8]. In our institution we routinely image the kidneys in an oblique coronal plane to outline their craniocaudad position and to define invasion of adjacent organs (e.g., liver, pancreas, adrenals) which lie either craniad to caudad to the lesion. The axial plane allows definition of tumor invasion into the perirenal space, involvement of psoas muscle and lymph nodes. Sagittal and oblique planes are optional and can be diagnostic, depending on the location and direction of tumor growth.

Pulse Sequences

MRI tissue contrast depends on differences in signal intensity in at least one of the sequences that are routinely employed. Here again early reports

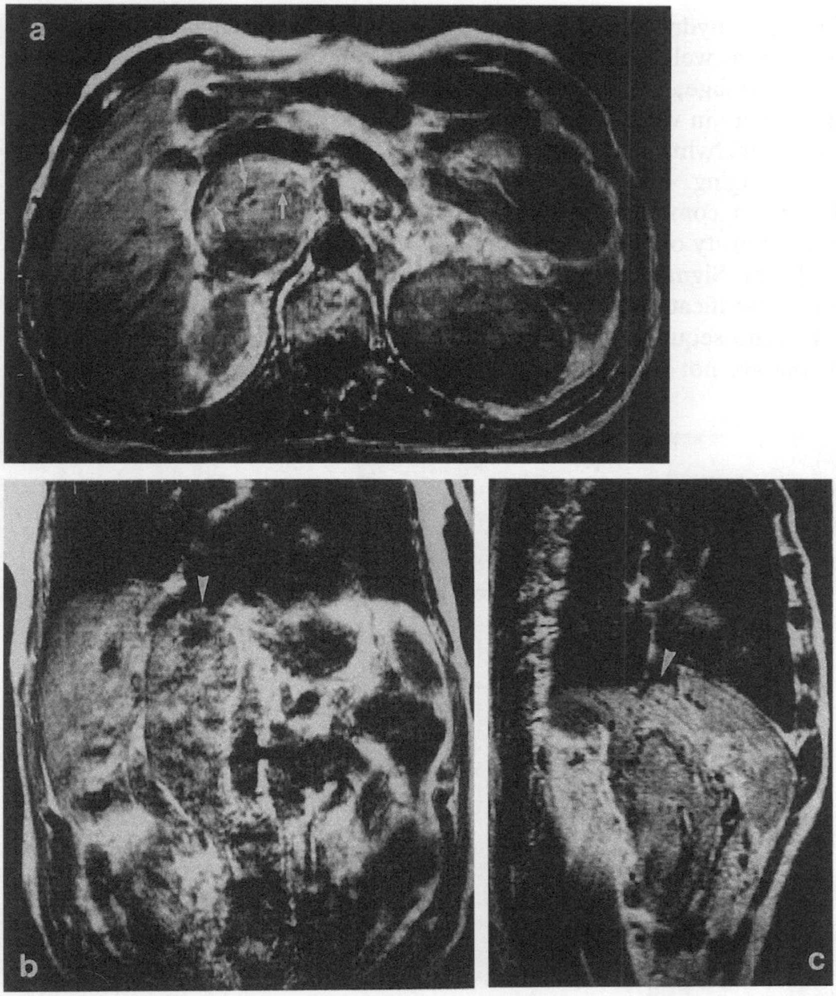

Fig. 1a–c. Extensive vena caval tumor thrombus in a patient with renal cell carcinoma. The axial scan (**a**) shows the left renal tumor hypointense due to hemorrhage and a caval thrombus of almost equal size with intratumoral vessels (*arrows*). Coronal (**b**) and sagittal (**c**) scans exactly define tumor extension up to the level of the hepatic veins (*arrowheads*). (All images T1-weighted: 580/20 ms)

claimed differences in either T1 or T2-weighted imaging techniques until an increasing number of tumors was found to be isointense to renal parenchyma in all pulse sequences [1, 9]. Despite occasional case reports MR signal intensity values and relaxation times of normal renal tissue and renal tumors do overlap [5]. In addition, renal tumors – not even a pathological entity as is RCC – lack a predictable signal pattern. Depending on the proton density, the degree of hydration or edema, the presence or absence of hemorrhage and necrosis as well as the time interval between these events and acquisition of the MR image, relaxation times for these tissues vary, and so does the signal intensity in various pulse sequences. Reactive changes of the adjacent normal parenchyma may lead to overestimation of tumor size and in some cases overstaging.

The most common appearance described in the literature is an intermediate intensity on T1-weighted images and hyperintensity on T2-weighted images [4, 5]. Signal intensity on spin-echo images alone, however, does not allow the classification of benign and malignant lesions; reports on the use of gradient-echo sequences in patients with solid renal tumors show improved results but are not yet convincing [5]. Some clinical data have suggested that

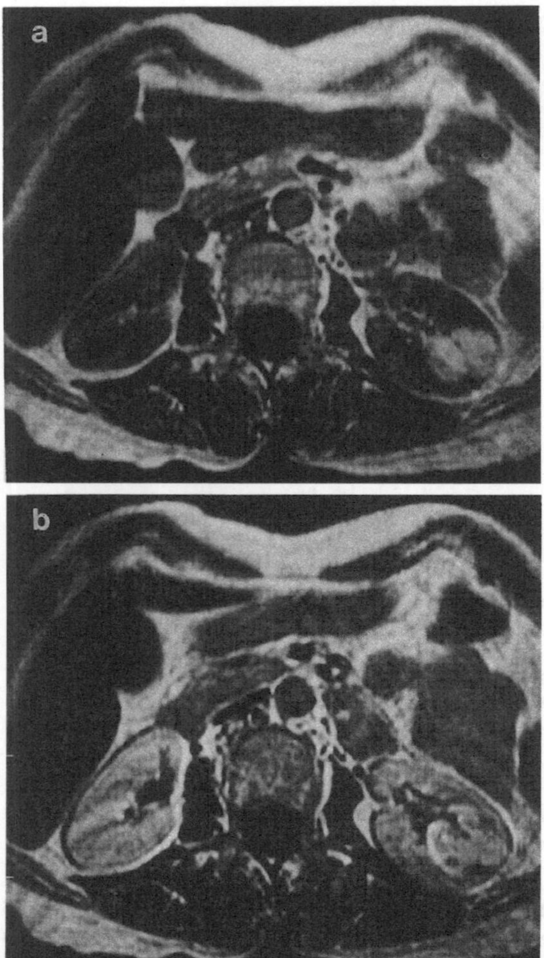

Fig. 2a,b. Axial T1-weighted images (600/20 ms) at the level of the left renal hilum. Before (**a**) and after (**b**) intravenous administration of 0.1 mmol/kg gadolinium-labeled diethylene-triaminopentaacetic acid. Note the high signal intensity of the left renal tumor, which is comparable to fat and the low intensity regions after contrast enhancement: typical appearance of an angiomyolipoma.

signal differences between tumor and normal parenchyma decreases with decreasing tumor size. A very well-designed and performed study by Quint and coworkers [7], however, confirmed that small tumors that were not detected on various in vivo pulse sequences showed the same signal pattern as large tumors when imaged in vitro with surface coils immediately after resection. It thus seems that motion artifacts and partial volume effects of larger coils may blur the tissue contrast and thereby obviate a definite distinction between normal and tumorous tissue. Here, again, fast imaging techniques using breath-hold technique may solve this diagnostic problem. The diagnosis of small RCC will until then rely on indirect signs of tumor growth: contour deformities and interruption of the corticomedullary border.

The small group of tumors deriving from mesenchymal tissues shows quite characteristic features on MR images [6], depending on the prevalent tissue type (Fig. 2). Large vessels in angiomatous lesions are easily detected by their typical contour and luminal loss of signal in routine pulse sequences. Lipomatous tissues gives hyperintense signals on T_1-weighted images. Tumors consisting mainly of connective tissue are sometimes hard to distinguish from chronic inflammatory change or scar formation, although their contour often permits differentation. Due to the low incidence of these tumors the literature offers no definite numbers for sensitivity, specificity, or accuracy of MRI.

Intravenous Contrast

Experience with the use of gadolinium-labeled diethylenetriaminopentaacetic acid in renal tumors is very limited. In a series of 44 patients Fein et al. found no definite pattern of contrast enhancement [1]. The results can be well explained by the heterogeneous tissue composition of these tumors, caused by necrosis, hemorrage, and extremely variable vascularization.

Conclusions

Results of MRI in detection and staging of solid renal tumors have been encouraging but have failed to surpass the established imaging methods, especially dynamic CT. With the use of fast imaging techniques the spatial resolution due to elimination of respiration artifacts is improving. Contrast agents improve the diagnostic value if tumor and normal renal parenchyma differ in vascularity. The undisputed advantages of MRI are its multiplanar imaging capability and its high accuracy to define vascular, especially venous involvement in malignant disease. For imaging of renal masses MRI should be used when other imaging modalities are equivocal, and when patients with a history of renal failure or severe adverse reactions cannot be exposed to iodinated contrast agents.

References

1. Fein AB, Lee JKT, Balfe DM, Heiken JP, Ling D, Glazer HS, McClennan BL (1987) Diagnosis and staging of renal cell carcinoma: a comparison of MR imaging and CT. AJR Am J Roentgenol 148:749–753
2. Horan JJ, Robertson CN, Choyke PL, Frank JA, Miller DL, Pass HI, Linehan WM (1989) The detection of renal carcinoma extension into the renal vein and inferior vena cava: a prospective comparison of venocavography and magnetic resonance imaging. J Urol 142:943–948
3. Hricak H, Demas BE, Williams RD, McNamara MT, Hedgcock MW, Amparo EC, Tanagho EA (1985) Magnetic resonance imaging in the diagnosis and staging of renal and perirenal neoplasms. Radiology 154:709–715
4. Hricak H, Thoeni RF, Carroll PR, Demas BE, Marotti M, Tanagho EA (1988) Detection and staging of renal neoplasms: a reassessment of MR imaging. Radiology 166:643–649
5. Krestien GP (1991) Morphological and functional MR of the kidneys and adrenal glands. Field and Wood, New York
6. Kulkarni MV, Shaff MI, Sandler MP, Tishler J, Winfield AC, Patton JA, Wolfe O, Partain CL, James AE Jr (1984) Evaluation of renal masses by MR imaging. J Comput Assist Tomogr 8:861–865
7. Quint LE, Glazer GM, Chenevert TL (1988) In vivo and in vitro MR imaging of renal tumors: histopathologic correlation and pulse sequence optimation. Radiology 169: 359–362
8. von Schulthess GK (1989) Morphology and function in MRI. Springer, Berlin Heidelberg New York
9. Uhlenbrock D, Fischer C, Ruehl G, Beyer HK, Hummelsheim P (1987) Kernspintomographie und Computertomographie des malignen Hypernephroms. Fortschr Röntgenstr 146:664–674

Springer-Verlag and the Environment

We at Springer-Verlag firmly believe that an international science publisher has a special obligation to the environment, and our corporate policies consistently reflect this conviction.

We also expect our business partners – paper mills, printers, packaging manufacturers, etc. – to commit themselves to using environmentally friendly materials and production processes.

The paper in this book is made from low- or no-chlorine pulp and is acid free, in conformance with international standards for paper permanency.

MIX
Papier aus verantwortungsvollen Quellen
Paper from responsible sources
FSC® C105338

If you have any concerns about our products,
you can contact us on
ProductSafety@springernature.com

In case Publisher is established outside the EU,
the EU authorized representative is:
Springer Nature Customer Service Center GmbH
Europaplatz 3, 69115 Heidelberg, Germany

Printed by Libri Plureos GmbH
in Hamburg, Germany